T0282621

CAMBRIDGE LIBRARY COLLECTION

Books of enduring scholarly value

Earth Sciences

In the nineteenth century, geology emerged as a distinct academic discipline. It pointed the way towards the theory of evolution, as scientists including Gideon Mantell, Adam Sedgwick, Charles Lyell and Roderick Murchison began to use the evidence of minerals, rock formations and fossils to demonstrate that the earth was older by millions of years than the conventional, Bible-based wisdom had supposed. They argued convincingly that the climate, flora and fauna of the distant past could be deduced from geological evidence. Volcanic activity, the formation of mountains, and the action of glaciers and rivers, tides and ocean currents also became better understood. This series includes landmark publications by pioneers of the modern earth sciences, who advanced the scientific understanding of our planet and the processes by which it is constantly re-shaped.

The Fossil Flora of Great Britain

Employed early on in his career by Sir Joseph Banks, the botanist John Lindley (1799–1865) went on to conduct important research on the orchid family and also recommended that Kew Gardens should become a national botanical institution. This pioneering three-volume work of palaeobotany, first published between 1831 and 1837, catalogues almost 300 species of fossil plants from the Pleistocene to the Carboniferous period. The geologist and palaeontologist William Hutton (1797–1860), with whom Lindley collaborated, was responsible for collecting the fossil specimens from which the 230 plates were drawn. The first serious attempt at organising and interpreting the evidence of Britain's primeval plant life, this resource is notable also for its prefatory discussion of topics such as coal seams and prehistoric climate. Volume 1 opens with a context-setting introduction and list of genera, followed by the descriptions of plates 1-79.

Cambridge University Press has long been a pioneer in the reissuing of out-of-print titles from its own backlist, producing digital reprints of books that are still sought after by scholars and students but could not be reprinted economically using traditional technology. The Cambridge Library Collection extends this activity to a wider range of books which are still of importance to researchers and professionals, either for the source material they contain, or as landmarks in the history of their academic discipline.

Drawing from the world-renowned collections in the Cambridge University Library and other partner libraries, and guided by the advice of experts in each subject area, Cambridge University Press is using state-of-the-art scanning machines in its own Printing House to capture the content of each book selected for inclusion. The files are processed to give a consistently clear, crisp image, and the books finished to the high quality standard for which the Press is recognised around the world. The latest print-on-demand technology ensures that the books will remain available indefinitely, and that orders for single or multiple copies can quickly be supplied.

The Cambridge Library Collection brings back to life books of enduring scholarly value (including out-of-copyright works originally issued by other publishers) across a wide range of disciplines in the humanities and social sciences and in science and technology.

The Fossil Flora of Great Britain

*Or, Figures and Descriptions of the Vegetable
Remains Found in a Fossil State in this Country*

VOLUME 1

JOHN LINDLEY
WILLIAM HUTTON

CAMBRIDGE
UNIVERSITY PRESS

CAMBRIDGE
UNIVERSITY PRESS

University Printing House, Cambridge, CB2 8BS, United Kingdom

Published in the United States of America by Cambridge University Press, New York

Cambridge University Press is part of the University of Cambridge.
It furthers the University's mission by disseminating knowledge in the pursuit of
education, learning and research at the highest international levels of excellence.

www.cambridge.org
Information on this title: www.cambridge.org/9781108068543

© in this compilation Cambridge University Press 2014

This edition first published 1831–3
This digitally printed version 2014

ISBN 978-1-108-06854-3 Paperback

THE

FOSSIL FLORA

OF

GREAT BRITAIN;

OR,

FIGURES AND DESCRIPTIONS

OF THE

VEGETABLE REMAINS FOUND IN A FOSSIL STATE

IN THIS COUNTRY.

BY

JOHN LINDLEY, Ph. D. &c. &c.

PROFESSOR OF BOTANY IN THT UNIVERSITY OF LONDON;

AND

WILLIAM HUTTON, F.G.S. &c.

_____

" Avant de donner un libre cours à notre imagination, il est essentiel de rassembler un plus grand nombre de faits incontestables, dont les conséquences puissent se déduire d'elles-mêmes."—*Sternberg.*

_____

VOLUME I.

_____

LONDON:

JAMES RIDGWAY, PICCADILLY.

—

1831-3.

Tilling, Printer, Chelsea.

PREFACE.

THE importance of Organic Remains in pointing out the changes which the surface of the Globe has undergone, during periods beyond the reach of traditionary record, has long been acknowledged.

By their assistance, we are enabled to snatch a glimpse of the early history and condition of our Planet, and of the successive races of organized bodies which have existed upon it. In fact, a very large part of Modern Geology is founded upon the evidence which they afford.

Whilst this has been generally confessed, it was chiefly to the remains of animals that Naturalists, for a long time, directed their attention; although there are many questions of deep interest in the elucidation of the History of the Globe, which are likely to be solved by the study of the position held in the Vegetable Kingdom, by plants now known only in a fossil state.

b

The identity of certain strata in which few animal remains are now to be discovered—the probable condition of the atmosphere at the most remote periods—what gradual changes that climate may have undergone since living things first began to exist—whether there has been, from the commencement, a progressive development of their organization—all these are questions which it is either the peculiar province of the Botanist to determine, or which his enquiries must, at least, tend very much to elucidate.

Considerations of this kind have gradually forced themselves upon the minds of Geologists, until the overcoming the difficulties that offer themselves to a strict examination of fossil vegetable remains has come to be an object of indispensible necessity. It is found, that neither a barren nomenclature, destitute of all attempt at determining the relations that former species bore to those of our own æra, nor supposed identifications of species by vague external characters, nor hasty determinations of analogies by means of partial views of structure, are sufficient to satisfy the geological enquirer; on the contrary, it is now distinctly seen, that nothing short of a most rigorous examination is likely to serve the ends of science, and that all conclusions that are not drawn from the most precise evidence that the nature of the subject will afford, must either be rejected, or, at least, received with the greatest caution.

Unfortunately, Fossil Botany is beset with difficulties of a peculiar character. The materials that the enquirer has to work upon, are not only disfigured by those accidents to which all fossil remains are exposed in common, but they are also those which would, in recent vegetation, be considered of the smallest degree of importance. There is, in most cases, an almost total want of that evidence by which the Botanist is guided in the examination of recent plants; and not only the total destruction of the parts of fructification, and of the internal organization of the stem, but what contributes still more to the perplexity of the subject, a frequent separation of one part from another, of leaves from branches, of branches from trunks, and if fructification be present, of even it from the parts of the plant on which it grew, so that no man can tell how to collect the fragments that remain into a perfect whole. For it must be remembered, that it is not in Botany, as in Zoology, where a skilful anatomist has no difficulty in combining the scattered bones of a broken skeleton. In Botany, on the contrary, the component parts of both foliage and fructification are often so much alike in outline, which is all that the Fossil Botanist can judge from, as to indicate almost nothing when separated from each other, and from the axis to which they appertain. It is only by the various combinations of these parts that the genera and species of plants are to be recognized, and it is precisely these combinations that in fossils are destroyed.

Insurmountable as these obstacles may, at first sight, appear, it must be confessed that they have yielded, in a degree that could scarcely have been anticipated, to the persevering investigations of a few skilful observers, who, combining great acuteness with all the power that the modern state of Botanical science can afford them, have clearly pointed out the possibility of reading one of the darkest, but most interesting pages in the history of the globe. The æra of Sternberg, Martius, Buckland, Witham, and more especially of Adolphe Brongniart, will be that from which future Geologists will date the origin of Fossil Botany, as a separate branch of science. The latter of these writers, in particular, has embodied what is at present known of the subject in a work, which, independently of its other merits, may fairly lay claim to being by far the most extensive, and best arranged general treatise upon the ancient vegetation of the world. For ourselves, notwithstanding the many points in which we find it necessary to differ in opinion, we have no hesitation in recommending M. Brongniart's book to Geologists, as the most safe guide they can follow in all that relates to Fossil Botany.

Having stated thus much, we might, on the present occasion, content ourselves with a bare explanation of the objects we have in view, in the

work that is now laid before the public, or, at
least, with touching upon such points only as
concern the elucidation of the more immediate
object of our enquiry, by the discovery of those
lost characters of species, which, no doubt, are
still locked up in our mines and rocks, whence
it is to be hoped the skilful observer will, in
time, extract them. But, as the whole subject
is one of great interest, and as it is impossible to
say to what future discoveries may lead, we beg
leave to offer a few brief observations upon the
existing state of what is known or conjectured, in
regard to Fossil Botany, and especially upon some
of those topics, which being of the most striking
importance, are those with regard to which it is
more particularly desirable that exact information
should be obtained.

That the face of the globe has successively undergone total changes, at different remote epochs,
is now a fact beyond all dispute; as, also, that
long anterior to the creation of man, this world
was inhabited by races of animals, to which no
parallels are now to be found; and that those
animals themselves only made their appearance
after the lapse of ages, during which no warm
blooded creatures had an existence. It has been
further remarked by Zoologists, that the animals
which first appeared in these latitudes, were
analogous to such as now inhabit tropical regions
exclusively; and that it was only at a period im-

mediately antecedent to the creation of the human race, that species, similar to those of the existing æra, began to appear in northern latitudes.

Similar peculiarities have been also found to mark the vegetation of corresponding periods. It would hardly be credited, by persons unacquainted with the evidence upon which such facts repose, that, in the most dreary and desolate northern regions of the present day, there once flourished groves of Tropical plants, of Coniferæ like the Norfolk Island and Araucarian Pines, of Bananas, Tree-ferns, huge Cacti, and Palms; that the marshes were filled with rush-like plants, fifteen or twenty feet high, the coverts with ferns like the undergrowth of a West Indian Island, and that this vegetation, thus inconceivably rich and luxuriant, grew amidst an atmosphere that would have been fatal to the animal world. Yet, nothing can well be more certain than that such a description is far from being overcharged. In the *Coal formation*, which may be considered the earliest in which the remains of land plants have been discovered, the Flora of England consisted of ferns, in amazing abundance, of large Coniferous trees, of species resembling Lycopodiaceæ, but of most gigantic dimensions, of vast quantities of a tribe, apparently analogous to Cacteæ, or Euphorbiaceæ, but, perhaps, not identical with them, of Palms, and other Monocotyledones; and, finally, of numerous plants, the exact nature of which is as yet extremely doubtful. Between two and three hundred species have been detected in this formation, of which two-thirds are ferns.

In the *New Red Sand-stone formation*, the characters of vegetation appear to be altered by the disappearance of the gigantic Cacteæ, or Euphorbiaceæ, by a diminution of the proportion of Ferns, and by the appearance of a few new tribes; but so little is yet known of the Flora of this period, that it is scarcely worth taking it into account.

In the *Lias* and *Oolitic formations*, an entirely new race of plants covered the earth. The proportional number of Ferns is diminished, the gigantic Lycopodium-like and Cactoid plants of the Coal Measures, Calamites, and Palms, all disappear; vegetation has no longer a character of excessive luxuriance, but species, undoubtedly belonging to Cycadeæ, and analogous to plants, now natives of the Cape of Good Hope, and of New Holland, appear to have been common; Coniferous plants were still plentiful, but they were of species that did not exist at an earlier period. Whether any other Dicotyledons, than those of the Cycas and Pine tribes, existed at this time, does not clearly appear.

Up to this time, the features of vegetation were exclusively extra-European, and chiefly tropical; but immediately succeeding *the Chalk*, a great change occurred, and a decided approach to the Flora of modern days took place in some striking particulars. The *Plastic Clay* formation is cha-

b 4

racterized by a total absence of Cycadeæ, the number of Ferns is again diminished, Coniferæ increase in quantity, and, mixed with Palms and other Tropical Monocotyledons, there grew Elms, Willows, Poplars, Chesnuts, and Sycamores, along with multitudes of Dicotyledonous plants, not at present determined.

But little remains of the vegetation of succeeding periods; but this little suffices to shew, that a gradual change to the existing state of things was still in progress. In the *Lower Fresh Water formation*, one species of Palm still maintained an existence, and it would seem, that it was accompanied by a few Tropical Trees, such as Cecropia, Sterculia, and some Malvaceæ.

Finally, in the *Upper Fresh Water formation*, nothing has been found to distinguish the Flora from that of the present day, except in regard to species.*

Such are the conclusions to which Geologists have arrived, from an examination of the data that actually exist; but it must be confessed, that however important the evidence already procured may undoubtedly be considered, it is as nothing compared with what is to be expected from future discoveries. At the period of the Coal formation, when vegetation was far more copious than it is in any part of the world, at the present day, less than 300 species are known; and M. Brongniart enu-

---

* In all these statements, Marine Plants have been intentionally omitted.

merates only 19 from the New Red Sandstone, 74 from the Lias and Oolitic beds, 34 from the lowest Tertiary Rocks, and 54 from all the superior strata. The field of Fossil Botany may, therefore, be said to be scarcely entered, and the materials hitherto accumulated must be understood as bearing a very small proportion to those still remaining to be discovered. Hence, calculations of the proportion borne by one tribe to another, in a given formation, are by no means to be depended upon ; for the discovery of a very few additional species may, where such inconsiderable numbers are concerned, entirely alter the result.* Still less can we be justified in assuming, that certain races of plants had no existence at any former period ; thus, the Coniferous tribe, which was, in 1828, excluded by M. Adolphe Brongniart from the Coal formation, has now been demonstrated to exist there in great abundance, and, in some cases, in a state closely approaching that of modern times. *(See plates* 1, 2, 3, 23, 24, *of this work.)* In further illustration of the same remark, it may be observed, that no trace of any glumaceous plant has been met with, even in the latest Tertiary Rocks, although we know that Grasses now form a portion, and, usually, a very

* This is already apparent from the additions made by Messrs. Phillips, and Bird and Young, to the Oolitic Flora ; additions, however, of which we have not been able to avail ourselves in computing the number of the Flora, because it is impossible to tell what of their new species are different from those named, but not defined by Adolphe Brongniart.

considerable one of every Flora of the world, from
New South Shetland, to Melville Island, inclu-
sive. It may, indeed, be conjectured, that before
the creation of herbivorous animals, Grasses and
Sedges were not required, and, therefore, are not
to be expected in any beds below the Forest
Marble, and Stonesfield Slate; but it is difficult
to conceive how the animals of the upper Tertiary
beds could have been fed, if Grasses had not then
been present.

That the temperature of this climate was, in
the beginning, that of the Tropics, is legitimately
inferred from the nature of the vegetation of the
coal measures, as compared with that of the present day; for it is found, that the large proportion
there borne by Ferns to other plants, is now a
characteristic only of certain Tropical Islands.
The existence of Palms is a corroboration, although not, in itself alone, a sufficient evidence of
the same fact; and the great dimensions of certain
plants, such as Sigillariæ, the exact nature of
which is uncertain, but which seem most analogous to Cacteæ, or Euphorbiaceæ, together with
the presence of Stigmaria, are all additional proofs
of a high temperature, accompanied by great
atmospheric humidity. It is curious, nevertheless,
to remark the questionable nature of the evidence,
popularly adduced in proof of a former tropical
climate in England; viz. the existence of gigantic Tree Ferns and Palms in the Coal Mines.
The latter plants are found now in the South of
Europe, and in Barbary; and it may, therefore, be

supposed that a very moderate elevation of temperature, by no means Tropical, would enable them to grow in more northern latitudes ; and, besides, they are so uncommon in the Coal measures, that three species only have been discovered, and those are of very rare occurrence ; while of all the supposed species of Tree Fern Stems, enumerated by M. Brongniart, Count Sternberg, and others, under the name of Sigillaria, or its synonyms, there is, in all probability, not one that can be botanically recognized as such. A single specimen of a Tree Fern Stem, in the Coal measures, has been pointed out to us by our friend Mr. Lonsdale ; but we know of no other instance.

Connected with this subject is a circumstance that we do not remember to have seen adverted to ; but which, nevertheless, appears to us to form one of the most curious problems that the philosopher has yet to solve. It is well known that numerous remains of Mastodons were found in Melville Island : now, what kind of plants, fit for the food of such monstrous animals, could at any period, during which the axis of the world remained in its present direction, have possibly grown in such latitudes, let the temperature have been what it might ? These animals must have had plants in abundance to live upon, in a country which, at the present day, affords so little means of vegetation, that the largest tree is a Willow, six inches high. It has been said, that if it be allowed that, in former ages, central fires operated conjointly

with the same solar heat as exists at the present
day, that cause alone may have been sufficient to
have produced a tropical atmosphere in such countries as Great Britain, without any change in the
axis of the earth ; a postulate, which, we think, may
be safely granted. But it seems to have been
overlooked, that this cannot also be conceded in
regard to climates like that of Melville Island ;
because, supposing the axis of the earth to have
been always the same, that spot must, necessarily,
from its polar situation, have been always, for many
months in every year, in darkness ; a condition
under which no plants can exist, at the present
day, unless in a torpid state. But if we can judge
of the ancient vegetation of Melville Island, by
that of Baffin's Bay, it was very like that of Great
Britain at the time of the Coal formation ; and
this was surely a vegetation in which there was
no torpidity, and to which the bright light, as well
as the high temperature of the tropics, must have
been indispensible. And although the æra of the
Mastodons, in Melville Island, is much more recent than that of the Coal measures, yet we are
justified in assuming, that if the vegetation of a still
more remote period was not calculated to develope
under a long absence of light, neither was that of
the æra of the Mastodons. Unless this difficulty
can be explained, which, we think, is possible, the
state of vegetation about the north pole, in former
times, can only be accounted for by a difference in
the direction of the axis of the earth ; for light is

an agent, without which no growing plants can exist, at the present day, for a single week, even in a low temperature, without suffering serious injury.

Of a still more questionable character is the theory of *progressive development*, as applied to the state of vegetation in successive ages. The opinion, that in the beginning, only the most simple forms of animals and plants were created, and that, in succeeding periods, a gradual advance took place in their degree of organization, till it was closed by the final creation of warm blooded animals, on the one hand, and of Dicotyledonous Trees, on the other, is one that very generally prevails. How far this may be admissible in the animal world, it is for Zoologists to determine; but, in the Vegetable Kingdom, it cannot be conceded, that any satisfactory evidence has yet been produced upon the subject; on the contrary, the few data that exist, appear to prove exactly the contrary. It is, therefore, very remarkable, that M. Adolphe Brongniart should adopt this view, and still more so, that one of his critics, an anonymous, but evidently very acute Geologist, should declare, that " the law of the progressive deve-
" lopment of the classes of plants, and of a
" gradual perfection of their organization, from
" the remotest periods, till the latest geological
" epoch, is proved by this investigation, in as
" striking and evident a manner as has been done
" among the incomparably more numerous tribes

" of the animal kingdom, belonging to a former
" age."* The ground of this opinion is, that no
Dicotyledonous plants existed at the period of the
Coal formation, but that vegetation was, at that
time, composed of Cryptogamic, and Monocotyledonous plants alone.

With reference to this subject, we would, in the
first place, ask, what trace is there of the simplest
forms of Flowerless vegetation in the Coal measures, such as Fungi, Lichens, Hepaticæ, or
Mosses? to say nothing of Confervæ; many of
these would have communicated their casts as
distinctly to the matter that enveloped them, as
Ferns and Lycopodiaceæ, had they existed ; but
no trace of them is found ; we have, on the contrary, in their room, the most perfectly organized
plants of the Flowerless or Cryptogamic class,
namely, Ferns, Lycopodiaceæ, and supposed Equisetaceæ. Secondly, we are told that of Monocotyledones, the remains consist of Palms and plants
apparently analogous to Dracænas, Bananas, and
the Arrow Root tribe (Marantaceæ); but are
these plants of imperfect organization? either considered *per se*, or when compared with the rest of
the class to which they appertain ; on the contrary, they are the most highly developed tribes
that are known in the Monocotyledonous class
of the existing æra; the simplest forms of Mono-

* Edinburgh New Philosophical Journal, October, 1829,
p. 112.

cotyledonous vegetation, are Grasses, Sedges, Rushes, Fluviales, and other plants related to Aroideæ, of none of which is there any the slightest trace anterior to our own æra, unless a Phyllites multinervis, and a few other plants in the green-sand, and beds above the chalk, should prove to belong to one of them. But, it is said, there are no Dicotyledonous plants in the Coal measures; we pass by the fact that has now been so well ascertained, that Coniferous trees were abundant at the period of the Coal formation, because an argument might be raised about the *dignity* of Coniferæ, among Dicotyledonous plants; but what were Sigillariæ, or at least Stigmariæ, the latter of which must have been one of the most common genera of the period, if we may judge from the thousands of fragments that still remain; that the former were Tree Fern stems, as is generally supposed, seems to us in the highest degree improbable, as we hope hereafter to explain; that the latter were not Lycopodiaceæ, we trust, we have demonstrated already; the weight of evidence seems to incline in favour of both having been Dicotyledonous plants, and of the highest degree of organization, such as Cacteæ, or Euphorbiaceæ, or even Asclepiadeæ; at least, there is nothing whatever to prove the contrary. The result of this investigation is well worthy of attention; it shews that, so far from " a gradual perfection of organization having been " going on from the remotest period, till the latest

"geological epoch," some of the most perfect
forms of each of the three great classes of the
Vegetable Kingdom were among the very first
created; and that either the more simple plants
of each class did not appear till our own æra, or
that no trace of them at an earlier period has been
preserved. But, supposing that Sigillarias and
Stigmarias could really be shewn to be Cryptogamic plants, and that it could be absolutely
demonstrated, that neither Coniferæ nor any other
Dicotyledonous plants existed in the first Geological age of land plants, still the theory of progressive development would be untenable, because it would be necessary to shew, that Monocotyledons are inferior in dignity, or, to use a more
intelligible expression, are less perfectly formed
than Dicotyledons. So far is this from being the
case, that if the exact equality of the two classes
were not admitted, it would be a question whether Monocotyledons are not the more highly
organized of the two; whether Palms are not of
greater dignity than Oaks, and Cerealia than
Nettles.

In looking at the general character of the successive periods of ancient vegetation, we cannot
fail to be struck with the greater variety of Fossil
species in the oldest, than in the newest rocks;
and that as far as discoveries have gone, it would
appear, as if the number of species that have been
preserved, was in proportion to the antiquity of
the formation. Thus, in Brongniart's Prodromus,

omitting his Transition formation, the plants of
which seem to belong rather to the Coal measures;
we find in the latter 258 species of land plants
enumerated; in the New Red Sandstone only 19;
in the variegated Marles and Lias 22 ; in the
Oolitic series 49; in the Plastic Clay formation
35; in the London Clay 16; in the Lower Fresh
Water formation 15; and in the Upper Fresh
Water formation 6 species.

This is certainly not owing to any actual paucity
of species in those periods of which the fewest
remains have been preserved, but to some cause
which protected the more ancient remains from
destruction by the atmosphere, and prevented the
carbon fixed in them from being lost. Much probability is attached to the conjecture of M.
Adolphe Brongniart, that the atmosphere, at
the time of primitive vegetation, was far more
charged with carbonic acid gas than now, and that
it was this which not only enabled gigantic species
to develope, at a time when there was little soil to
support them, but, also, in some measure, prevented their dead remains from being decomposed
by the action of the oxygen of the atmosphere.
It is further supposed, that the excess of carbon
thus assumed to have existed, which would have
been fatal to all air-breathing animals, was gradually abstracted from the atmosphere by plants,
until the air became fit, in the first place, for the
respiration of reptiles, and next, for that of Mam-
mālia. To the first part of this proposition there

c

is no botanical objection ; into the latter it is not our province to enquire.

Dismissing this part of our subject, we will next explain what the objects are of the work we have now commenced. We propose, in the first place, to combine, in a single point, figures of all the Fossil plants that have been discovered in the rocks of this country. The utility of such a work for *recent* plants, is attested by the English Botany of Mr. Sowerby ; and, no doubt, a similar publication upon our Fossil Flora will become, in time, a great mass of facts, to which Geologists will find it much more convenient to refer, than if the same information were scattered through many distinct publications. A similar object is, indeed, pursuing in France by M. Adolphe Brongniart, of whose Histoire des Végétaux Fossiles it is difficult to speak too highly ; but we confess that this, far from discouraging us in our own attempt, acts rather as a stimulus to greater exertion. Besides, we are not ashamed to confess that we have national feeling enough to make us anxious that the elucidation of every thing that relates to England, should come from the hands of Englishmen ; and that we should not be subject to the disgrace of being obliged to send our native Fossils to another country for examination, from want of the skill to determine them ourselves. The richness of Great Britain, in the Fossil remains of Vegetables, is well known to every Geologist ; and the facilities of studying them are

so great in the extensive excavations of our Coal
Mines, that it is in this country more especially
that information should be looked for upon the
subject.

In another point of view, we think a work of
this kind likely to be of general utility. It is a
very remarkable fact, that, in former ages, the range
of the species of plants was far more extensive
than at the present day. If we compare the
Floras of modern Europe and America, we find,
that they differ in the greater part of their species,
so that the general characters of the vegetation of
the two countries are now essentially unlike. But
M. Adolphe Brongniart assures us, that the plants
of the North American Coal Mines are, for the
most part, perfectly identical with those of
Europe, and that they all belong to the same
genera ; the same is stated of Fossils, from
Greenland, and from Baffin's Bay ;· that ours are
very much the same as those of the rest of
Europe, is also certain. A Fossil Flora of Great
Britain applies, then, not only to the rest of
Europe, as might have been expected, but also to
very distant countries.

In the third place, we hope, that a work
appearing periodically, may become the focus,
as it were, of all the knowledge that will be
gradually acquired in regard to this important
subject : that it will keep the enquiry in sight of
those, who, from their local position, will be able,
most powerfully, to aid it by the examination of

the remains within their reach ; but who may be the least acquainted with the nature of the information that is wanted, and with the progress that the science is making elsewhere. In order to render it as generally useful as our means will allow us,. we have added to these introductory remarks a concise arrangement of such genera of Fossil plants as are at present admitted, in which, perhaps, there is not much that is original, for it is necessarily based upon the work of M. Brongniart, so often already alluded to, but which serves to explain what our own views of the subject are, at the present moment, and contains such published additions as have been made since 1828. To each succeeding volume, we propose to prefix some similar table, either applied to genera or to species, corrected and made up to the time of its publication, by which the gradual advance of this branch of Geology will be made apparent ; and our work will constantly be upon a level with the existing state of science. We further propose to introduce, occasionally, lists, or even detailed accounts of the species found in particular localities, or formations ; so that, in this way, the local discoveries that may from time to time be made, will be constantly brought before the world. Many changes may be expected in a nomenclature, which is, at present, provisional to a great extent, and perhaps total alterations may take place in our ideas, respecting many fossils that have long been known. The biennial republication, now alluded to*, will be an effectual means of remedying the inconveniences that would otherwise attend such changes.

It must always be remarked, that, in this study, every one is a mere beginner; that he who has pursued it the longest, is still but upon the very threshold of the science, and that we have only just begun to clear away the impediments that accident and ages have accumulated in our path. It is no wonder that errors should be committed in such a pursuit. So perfectly hopeless is it to escape them, that Botanists have, probably, been deterred from engaging in the enquiry, as much by a dread of the risk to which their scientific reputation must necessarily be exposed, as by the difficulty of the task itself. For ourselves, however, we have no other object than the promotion of science, as far as our humble means will permit. We willingly place aside all considerations of personal loss of reputation, and we trust ourselves, if not fearlessly, at least cheerfully to the importance of our cause, to the aid and protection of those who can appreciate the peculiar nature of the enquiry; and to the persuasion, that unless, not one Botanist, but many, will lend their assistance to its elucidation, Geology must for ever remain deprived of the evidence to be afforded by

---

* As eight quarterly numbers of this work form a volume, the tables of genera or species will necessarily appear every two years.

a branch of science of almost equal importance with Zoology.

It will not be foreign to the object of these introductory observations, if we next proceed to explain, Firstly, in what way the state of the Fossil remains of plants renders it almost indispensible, that any investigation of their original structure should be conducted; and Secondly, what the chief points are, to which the attention of collectors should, more especially, be directed.

When a Botanist proceeds to the examination of a recent specimen of an unknown plant, he directs his view to certain peculiarities in the organs, both of fructification and vegetation, *taken together*; and from what he finds to be their structure, he judges of the class, order, or genus to which it belongs. But as in fossil plants neither calyx, corolla, stamens, nor pistillum, are to be recognized, an opinion has to be formed, not from the consideration of a complex combination of characters, in which the loss of one organ is compensated for by the peculiarities of those which remain : but from a few isolated and very imperfect data exclusively afforded by the remains of the organs of vegetation. In the latter, unfortunately, the modes of organization are not sufficiently varied, to enable us to draw any precise conclusions from their examination ; but, on the contrary, we are often obliged to be satisfied with a general idea only of the nature of the object of our enquiry. This is, perhaps, not attended with so much practical inconvenience as might be expected, in a Geological point of view, because the end of science will be sufficiently answered, if we can, in the first place, determine the general characters and affinities of the plants of former æras, and, in the second, so exactly classify their fossil remains, as to be able to recognize them, with such precision, as to render them available for the identification of strata.

It usually happens that the only parts which are capable of being examined in a fossil state, are the internal structure of the stem, and its external surface; together with the position, division, outline, and veining of the leaves. Of these it has never yet happened that any one specimen has afforded the whole; more frequently it is only two or three of those characters that the Botanist can employ.

Suppose that he has a fragment of the fossil *trunk* of some unknown tree; if no trace can be discovered of its exact anatomical structure, it may be possible, at least, to ascertain whether its wood was deposited in concentric zones, or in a confused manner; in the former case, it would have been Dicotyledonous, or Exogenous; in the latter, Monocotyledonous, or Endogenous; if a transverse section should shew the remains of sinuous unconnected layers, resembling arcs with their ends directed outwards, of a solid homogeneous character, and imbedded among some softer substance, then it may be considered certain that

such a stem belonged to some arborescent Fern.
But if the state of a fossil stem will admit of an
anatomical examination, it is always desirable
that it should be instituted with the assistance of
the microscope. Mr. Witham was the first to make
known the possibility of this being done; and if it
should prove that the condition of fossil remains
is in general favourable to this kind of examination, more light is likely to be thrown upon the
extinct Flora than could be otherwise anticipated.
If the tissue of a stem should be found entirely
cellular, and it could be satisfactorily made out,
that no vascular tissue whatever was combined
with it, the specimen would, in all probability,
have belonged to that division of the Vegetable
Kingdom, which, being propagated without the
agency of sexes, is by Botanists called Crypto-
gamia; a specimen of this kind should, however,
be examined with the most rigorous accuracy;
because it might have been a succulent portion
of some Dicotyledonous tree, in which the vascular system was so scattered among cellular substances as to be scarcely discernible.   If the
tissue should have consisted of tubes placed parallel with each other, without any trace of rays
passing from the centre to the circumference, it
would have been Endogenous, even if there
should be an appearance of concentric circles
in the wood; but if any trace whatever can be
discovered of tissue, crossing the longitudinal
tubes at right angles, from the centre to the circumference, then such a specimen would have

been Exogenous, whether concentric circles can
be made out or not; for such an arrangement of
tissue would indicate the presence of medullary
rays, which are the most certain sign of a Dicotyledonous plant. If, in a specimen having these
rays, the longitudinal tubes are all of the same
size, a circumstance obvious upon the inspection
of a tranverse section, the plant will have been
either Coniferous or Cycadeous; but, if among
the smaller tubes, which, in fact, are woody fibres,
some larger ones are interspersed *in a definite manner*; it would, in that case, have belonged to some
other tribe of Dicotyledons. It is indispensible that
the arrangement of the larger tubes should have
been definite, for appearances of the same kind
exist in much Coniferous wood; but, in the latter,
they are scattered in an indefinite manner among
the smaller tubes, and are not vessels, but cylindrical cavities for the collection of the resinous secre-
tion peculiar to the Fir tribe. Again, if the walls of
the longitudinal tubes of any fossil specimen are
found to exhibit appearances of little warts, grow-
ing from their sides, such a specimen had certainly belonged to some Coniferous or Cycadeous
plant, no other tribes whatever possessing such a
structure at the present day. Finally, if a trace
of pith can be discovered, that circumstance alone
will be a proof of the plant having been Dicotyledonous, because all other classes are destitute of
that central cellular column; it must, however, always be borne in mind, that absence of pith does

not prove that a specimen is not Dicotyledonous, because the roots of those plants have no pith.

If a stem is in such a state that nothing can be determined respecting its anatomy, we must then proceed to judge of it by another set of characters. In the first place, it should be enquired whether it had a distinctly separable bark, or a cortical integument that differed in its organization from the wood, without being separable from it; or neither the one nor the other. In the first instance, it would have been Dicotyledonous; in the second, Monocotyledonous ; in the third, Acotyledonous or Cryptogamic, supposing that it had been a trunk which many successive years had contributed to form. The distinction, as applied to the two latter classes, is not, however, so positive as could be wished, because Tree Ferns have a cortical integument ; but they are easily known by the long ragged scars left by their leaves ; and no other Cryptogamic plants possess the character of having a spurious bark. For this reason, it is doubtful whether Calamites is related to Equisetaceæ ; and if we could be sure that the coaly matter found enveloping that genus, was really the remains of a cortical integument, there would be no doubt of its affinity being of a different kind, as, for instance, with Juncus. But here is a difficulty; how are we to be sure, that this coaly matter is a part of the original organization of the stem, and that it is not an independent carbonaceous formation? Another object of enquiry will be, whether the

stem was articulated (as indicated by tumid *nodi*)
or not; and, if the former, whether it had the
property of disarticulating; these circumstances
are not of much positive value in pointing out
affinities ; but they afford negative evidence, that
must, on no account, be overlooked; for example,
if this had been properly considered in regard to
Calamites, although the affinity of that genus might
not have been discovered, yet it never could have
been referred either to Palms or Bambusas, which,
in no instance, ever disarticulate. A third, and
very important kind of evidence, is to be collected
from the scars left upon stems by the fall of leaves.
Although these will neither inform us of the
shape, or other characters of the leaves themselves, yet they indicate, with precision, their position, the form of their base, and sometimes,
also, their probable direction; we can tell, whether
they were opposite or verticillate, alternate or spirally disposed, deciduous or persistent, and imbricated or remote; all characters of great use, as
means of discrimination, and as often affording
important negative evidence upon doubtful points.
The Geologist will, however, be careful not to
ascribe too much value to modifications in the
origin of leaves, and, in particular, to the spiral
mode, which forms so striking a feature in many
Fossil remains; he will bear in mind that the
latter is theoretically the normal mode in which
all leaves originate, and that other modes are
more or less obvious modifications of it; and

finally, he will consider, that if he is not familiar
with instances of it in recent plants, it is because
the lines of spires are broken by the leaves that
are interposed between them and the eye. He
will, possibly, only remember, that the leaves of
Firs, the fruit of the Pine Apple, and the foliage
of the Screw Pine, (Pandanus) are arranged upon
this plan; but if he draws a line from base to
base of the leaves of any alternate-leaved plant,
always proceeding in the same direction, he will
find, that that line will describe a spire round the
axis from which the leaves originate; so that a
spiral appearance will be apparent in proportion
as leaves are approximated.

In judging of the identity of fossil stems, that
are characterized by their external appearance,
care must be taken not to distinguish as different
species those stems that have still their cortical
integument upon them, from such as have lost it.
In these two cases, the appearance of scars will
be different; those of the former being more
rounded, broader, and, probably, more deeply furrowed than the latter; for the one is a real scar,
shewing the outline of the base of the leaf, while
the latter is solely caused by the passage of
bundles of vessels out of the stem into the petiole
of the leaf.

The manner in which stems branch, is sometimes well deserving consideration; where no
trace of leaves can be found, their position may possibly be indicated by the origin of branches; for

the latter being always axillary to the leaves, can only originate as they do : but, unfortunately, the value of this fact is often reduced to nothing, by the appearance of branches from the axillæ of a few leaves only, in distant parts of the stem. The most useful character to be thus derived, is when the branches regularly bifurcate ; for this kind of ramification is a strong symptom of a cryptogamic plant, especially if accompanied by an imbricated foliage.

In *Leaves* we can rarely recognize, in a fossil state, more than their mode of *venation, division, arrangement,* and *outline,* to which are sometimes added their *texture* and *surface.* All these are of importance, but in unequal degrees. Of the highest value is the evidence afforded by the distribution of the veins, taken together with the mode of division of a leaf. If the veins are all parallel, unbranched, or only connected by little 'transverse bars, and the leaves undivided, the plant was probably Monocotyledonous ; and if the veins of such a leaf, instead of running side by side from the base to the apex, diverge from the midrib, and lose themselves in the margin, forming a close series of double curves; the plant was certainly analogous to what are now called Scitamineæ, Marantaceæ, and Musaceæ : but supposing that the parallel arrangement of simple veins is combined with a pinnated foliage, then the plant would probably have belonged to Cycadeæ, that curious tribe that stands on the very limits of

Monocotyledons, and Dicotyledons, and of Flowering and Flowerless plants. By such characters as these, however, there is no means of distinguishing certain Palms, if in a Fossil state, from Cycadeæ.

If veins are all of equal thickness, and dichotomous, we have an indication of the Fern tribe, which is seldom deceptive. Nevertheless, it must be remembered that the flabelliform leaves, both of Monocotyledons and Dicotyledons, have occasionally this kind of venation. Even if the veins are not dichotomous, if they are all of nearly equal thickness and very fine, or divided in a very simple manner, it is probable that they indicate the Fern Tribe, whether simple, as in the fossil genus, Tæniopteris; or reticulated, as in the modern genus Meniscium. If veins are of obviously unequal thickness, and so branched as to resemble the meshes of a net, we have a sign of Dicotyledonous structure that seldom misleads us. Finally, if no veins at all are to be found, an opinion must be formed, not from their absence, but from other circumstances. If the leaves are small, their absence may be due to incomplete development; but if the leaves are large and irregularly divided, we may have an indication of some kind of Marine plant. When leaves are small, and densely imbricated, they are generally considered, by Fossil Botanists, to belong to either Lycopodiaceæ, or Coniferæ; and there is so little to distinguish these families, in a fossil state, that there is scarcely any means of demonstrating to which such genera, as Lycopodites, Lepidodendron, Juniperites, Taxites, &c. and the like, actually belong.

It would be easy to extend these observations much further, but to dwell at length upon this branch of the subject, would carry us far beyond the limits of a preface. We will, therefore, bring our remarks to a conclusion, by calling attention to some of those points, to the elucidation of which, it is most to be wished, that Geologists, who have opportunities of collecting fossil plants, would apply themselves. In the first place, evidence is wanted as to the plants to which the cones called Lepidostrobi, the leaves called Lepidophylla, and the fruit named Cardiocarpa, respectively appertain; are they all portions of species of the same genus, or, as seems more probable, is not Cardiocarpon a part of a plant of a totally different affinity? Secondly, what were the leaves of Sigillaria, and of Stigmaria? Of the latter, something is known; but the leaves are always so crushed, that no notion can be formed of their exact nature. Mr. Steinhauer says, he has traced them to the length of 20 feet! In the third place, to determine the leaves of any of the fossil stems, that at present are only known in the latter state, such as Sternbergia, Bucklandia, Cycadeoidea, Caulopteris, Exogenites, and Endogenites, would be to supply a great desideratum. Again, what was the real nature of the stem of Calamites; was it an annual shoot proceeding from a perennial horizontal rhizoma, like that of Juncus, &c. ? Had it any leaves, and if so, were they of the nature of those figured in this work, as probably belonging to Calamites nodosus, but considered by Sternberg and Brongniart a distinct genus, which they call Volkmannia ? Another very interesting object of enquiry is into the anatomical structure of Lepidodendron, for the sake of settling whether that extensive fossil genus belonged to Coniferæ, or to Lycopodiaceæ, or to neither: We know nothing of the leaves belonging to the fossil fruits, called Anomocarpon, Musocarpon, &c. or of the fruit of Cycadeoidea, Annularia, Asterophyllites, and many others. Now these are difficulties that probably may be removed by diligent research among the beds in which such fossils occur ; and which, if removed, would contribute much more to fixing the science upon a solid basis, than the discovery of species not before described. For all such information as our friends may communicate upon these or similar subjects, we shall always make our grateful acknowledgments; and we trust, that when the time shall arrive for our laying before the world a further statement of the progress that Fossil Botany shall have made, we shall be able to announce that light has been thrown upon a part, at least, of those great questions, which are at present involved in great obscurity.

March 31, 1832.

GENERA OF FOSSIL PLANTS.

(March, 1832.)

N.B. Those genera marked (*) are recent; the remainder are only known in a fossil state. Characters are assigned to the latter chiefly.

Class 1. VASCULARES; or FLOWERING PLANTS.

Subclass 1. EXOGENÆ; or DICOTYLEDONS.

NYMPHÆACEÆ.†

Genus 1. * *Nymphæa.*
> One species—in the Upper freshwater formation.

LAURINEÆ.

Genus 2. * *Cinnamomum.*
> One species—in the Tertiary freshwater formation of Aix.

LEGUMINOSÆ.

Genus 2 a. *Phaseolites.* Leaves compound, unequally pinnate; leaflets entire, disarticulating, with nearly equal reticulated veins.
> One species—in the Tertiary freshwater formation of Aix.

ULMACEÆ.

Genus 3. * *Ulmus.*
> One species—in Tertiary formations.

CUPULIFERÆ.

Genus 4. * *Carpinus.*
> One species—in the Lignite of Tertiary beds.

Genus 5. * *Castanea.*
> One species—in Tertiary formations.

---

† The reader who is anxious for information regarding the characters of this, and the succeeding Natural Orders, is referred to the *Introduction to the Natural System of Botany.*

BETULINEÆ.

Genus 6. * *Betula.*

One species—in the Lignite of Tertiary beds.

SALICINEÆ.

Genus 7. * *Salix?*

One species—in Tertiary formations.

Genus 8. * *Populus.*

One or two species—in Tertiary formations.

MYRICEÆ.

Genus 9. * *Comptonia.*

One species—in the Lignite of Tertiary formations.

One species ?—in the Lower freshwater formation.

JUGLANDEÆ.

Genus 10. * *Juglans.*

Three species—in the Tertiary strata.

One species—in the upper bed of New red sand-stone.

EUPHORBIACEÆ.

? Genus 11. *Stigmaria.* (Variolaria *Sternb.* Mammillaria *Ad. Br.* Ficoidites *Artis.*) Stem originally succulent; marked externally by roundish tubercles, surrounded by a hollow, and arranged in a direction more or less spiral; having internally a distinct woody axis, which communicates with the tubercles by woody processes. Leaves arising from the tubercles, succulent, entire, and veinless, except in the centre, where there is some trace of a midrib.

Five or six species—in the Coal formation.

One species ?—in the Oolitic formation; viz. Mammillaria Desnoyersii of *Ad. Brongn. Ann. Sc.* 4. *t.* 19. *f.* 9, 10.

ACERINEÆ.

Genus 12. **Acer.*

One or two species—in the Tertiary beds.

CONIFERÆ.

† Wood only known.

Genus 13. *Pinites.* Axis composed of pith, wood in concentric circles, bark, and medullary rays, but with no vessels. Walls of the woody fibre reticulated.

Three species—in the Coal formation.

Genus 14. *Peuce.* Axis composed of pith, wood in concentric circles, bark, and medullary rays, but with no vessels. Walls of the woody fibre marked with oblong deciduous areolæ, having a circle in their middle.

One species—in the Coal formation.
Others—in the Oolitic formation.

†† Fruit, or branches and leaves, only known.

Genus 15. * *Pinus.* Leaves growing two, three, or five, in the same sheath. Cones composed of imbricated scales, which are enlarged at their apex into a rhomboidal disk. *Ad. Br.*

Nine species—in the Tertiary strata.

Genus 16. * *Abies.* Leaves solitary, inserted in eight rows in a double spire, often unequal in length, and distichous. Cones composed of scales, without a rhomboidal disk. *Ad. Br.*

One species.

Genus 17. *Taxites.* Leaves solitary, supported on a short petiole, articulated, and inserted in a single spire, not very dense, distichous. *Ad. Br.*

Five species—in the Tertiary beds.
One species—in the Oolitic formations.

Genus 18. * *Podocarpus.* Leaves solitary, much larger than in the last genus, sharp-pointed, flat, with a distinct midrib.

One species—in the Tertiary freshwater formation of Aix.

Genus 19. *Voltzia.* Branches pinnated. Leaves inserted all round the branches, sessile, slightly decurrent or dilated at the base, and almost conical; often distichous. Fruit forming spikes or loose cones, composed of distant imbricated scales, which are more or less deeply three-lobed. *Ad. Br.*

Four species—in the New red sandstone.

Genus 20. *Juniperites.* Branches arranged irregularly. Leaves short, obtuse, inserted by a broad base, opposite, decussate, and arranged in four rows. *Ad. Br.*

Three species—in the Tertiary beds.

Genus 21. *Cupressites.* Branches arranged irregularly. Leaves inserted spirally, in six or seven rows, sessile, enlarged at their base. Fruit consisting of peltate scales, marked with a conical protuberance in their centre. *Ad. Br.*

One species—in the New red sandstone.

Genus 22. * *Thuja.* Branches alternate, regularly arranged upon the same plane. Leaves opposite, decussate, in four rows. Fruit composed of a small number of imbricated scales, terminated by a disk, which has near its upper end a more or less acute, and sometimes recurved point. *Ad. Br.*

Three or four species—in the Tertiary formations.

Genus 23. *Thuytes.* Branches as in Thuja. Fruit unknown. *Ad. Br.*

Four? species—in schistose Oolite.

††† Doubtful Coniferæ.

Genus 24. *Brachyphyllum.* Branches pinnated, disposed on the same plane without regularity. Leaves very short, conical, almost like tubercles, arranged spirally. *Ad. Br.*

One species—in the lower Oolitic formation.

Genus 25. *Sphenophyllum.* (Rotularia *Sternb.*) Branches deeply furrowed. Leaves verticillate, wedge-shaped, with dichotomous veins.

Eight species—in the Coal formation.

CYCADEÆ.

† Leaves only known.

Genus 26. *Cycadites.* Leaves pinnated ; leaflets linear, entire, adhering by their whole base, having a single thick midrib ; no secondary veins. *Ad. Br.*

One species—in the Grey chalk.

Genus 27. * *Zamia.* Leaves pinnated; leaflets entire, or toothed at their extremity, pointed, sometimes enlarged and auricled as it were at their base, attached only by the midrib, which is often thickened; veins fine, equal, all parallel, or scarcely diverging. *Ad. Br.*

Fifteen species—in the Lias and Oolitic formation. One species—(bed unknown.)

Genus 28. *Pterophyllum.* Leaves pinnated; leaflets almost equally broad each way, inserted by the whole of their base, truncated at the summit; veins fine, equal, simple, but little marked, all parallel. *Ad. Br.*

Three species—in the Variegated marle of the Lias.
Three species—in the Sandstone of the Lias.
One species—in the Quadersandstein.
One species—in the lower Oolitic beds.

Genus 29. *Nilsonia.* Leaves pinnated; leaflets approximated, oblong, more or less elongated, rounded at the summit, adhering to the rachis by the whole of their base, with parallel veins, some of which are much more strongly marked than others. *Ad. Br.*

Two species—in the sandstone of the Lias.

†† Stems only known.

Genus 30. *Cycadeoidea.* Buckland. (Mantellia *Ad. Brong.*) Stem roundish or oblong, covered with densely imbricated scales, which are scarred at their apex.

Two species—in the Portland stone.

DICOTYLEDONOUS PLANTS OF DOUBTFUL AFFINITY.

? Genus 31. *Phyllotheca.* Stem simple, straight, articulated, surrounded at equal distances by sheaths, having long linear leaves, which have no distinct midrib.

One species—in the Coal formation.

Genus 32. *Annularia.* (Bornia *Sternb.*) Stem slender, articulated, with opposite branches springing from above the leaves. Leaves verticillate, flat, usually obtuse, with a single midrib, united at their base, of unequal length. *Ad. Br.*

Six or seven species—in the Coal formation.

Genus 33. *Asterophyllites.* (Bornia *Sternb.* Bruckmannia *Sternb.*) Stem scarcely tumid at the articula-

d 3

tions, branched. Leaves verticillate, linear, acute,
with a single midrib, quite distinct at their base.
(Fruit a one seeded? ovate, compressed nucule,
bordered by a membranous wing, and emarginate
at the apex. *Ad. Br.)*

Twelve species—in the Coal formation.
One species—in the transition beds.

Obs. This is probably an extremely heterogeneous assemblage, comprehending nearly all fossils with narrow veinless
verticillate leaves, that are not united in a cup at their base.

Genus 34. *Bechera.* Stem branched, jointed, tumid at the
articulations, deeply and widely furrowed. Leaves
verticillate, very narrow, acute, ribless?

One species—in the Coal formation.

Subclass 2. ENDOGENÆ; *or* MONOCOTYLEDONS.

MARANTACEÆ.

Genus 35. *Cannophyllites.* Leaves simple, entire, traversed
by a very strong midrib; veins oblique, simple,
parallel, all of equal size. *Ad. Br.*

One species—in a bed of coal, supposed to be more
recent than the old coal formation.

ASPHODELEÆ.

† Stems only known.

? Genus 36. *Bucklandia.* Stem covered by reticulated fibres,
giving rise to (imbricated) leaves which are not
amplexicaul, and the petioles of which are distinct to their base. *Ad. Br.*

One species—in Stonesfield slate.

Obs. Dr. Buckland suggests the possibility of this being
the amentum of a Cycadeous plant. *G. trans. vol.* 2. *n. s.*
p. 400.

Genus 37. *Clathraria.* Stem composed of an axis, the surface of which is covered by reticulated fibres,
and of a bark formed by the complete union of
the bases of petioles, whose insertion is rhomboidal. *Ad. Br.*

One species—in the Green sand?

†† Leaves only known.

? Genus 38. *Convallarites.* Leaves verticillate, linear, with parallel slightly marked veins. Stem straight, or curved. *Ad. Br.*

Two species—in the Variegated sandstone.

††† Flowers only known.

Genus 39. *Antholithes.*

One species—in the Tertiary beds.

***

SMILACEÆ.

Genus 40. *Smilacites.* Leaves heartshaped or hastate, with a well-defined midrib, and two or three secondary ribs on each side, parallel to the edge of the leaf. Veins reticulated. *Ad. Br.*

One species—in the Lower freshwater formation.

***

PALMÆ.

† Stems only known.

Genus 41. *Palmacites.* Stems cylindrical, simple, covered by the bases of petiolated leaves; petioles dilated, and amplexicaul. *Ad. Br.*

One species—in the lower beds of the London clay formation.

†† Leaves only known.

Genus 42. *Flabellaria.* Leaves petiolated, flabelliform, divided into linear lobes, plaited at their base. *Ad. Br.*

One species—in the Plastic clay formation.
One species—in the Lower freshwater formation.
One species—in the London clay formation.
One species—in the Coal formation.

Genus 43. *Phœnicites.* Leaves petiolated, pinnated; leaflets linear, united by pairs at the base, their veins fine, and little marked.

One species—in the Tertiary formations.

Genus 44. *Næggerathia.* Leaves petiolated, pinnated; leaflets obovate, nearly cuneiform, applied against the edges of the petiole, toothed towards their apex, with fine diverging veins. *Ad. Br.*

Two species—in the coal measures.

Genus 45. *Zeugophyllites.* Leaves petiolated, pinnated ; leaflets opposite, oblong or oval, entire, with a few
strongly marked ribs, confluent at the base and
summit, all of equal thickness. *Ad. Br.*

One species—in the Coal formation.

††† Fruit only known.

Genus 46. * *Cocos.* Fruit ovate, slightly three-cornered,
marked with three orifices near their base.

Three species—in the Tertiary formations.

––––––––

FLUVIALES.

Genus 47. *Zosterites.* Leaves oblong or linear, marked with
a small number of equal veins, which are at a
marked distance from each other, and are not
connected by transverse veins. *Ad. Br.*

Four species—in the Lower Greensand formation.
One species—in the Lias ?
Two species—in the Upper freshwater formation.

Genus 48. *Caulinites.* (Amphytoites *Desm.*) Stem branched,
bearing semi-annular, or nearly annular scars of
leaves, alternate in two opposite rows, marked
with little equal dots. *Ad. Br.*

One species—in the London clay formation.

––––––––

MONOCOTYLEDONOUS PLANTS OF DOUBTFUL AFFINITY.

† Stems only known.

Genus 49. *Endogenites.* This comprehends all fossil endogenous stems that do not belong to any of the
genera characterized separately. It is a mere
provisional assemblage of objects to be further
examined.

Several species—from the Tertiary strata.

Genus 50. *Culmites.* Stems articulated, with two or more
scars at the joints.

Three species—in the Tertiary beds.

Genus 51. *Sternbergia.* (Columnaria *Sternb.*) Stem taper,
slender, naked, cylindrical, terminating in a cone;
marked by transverse furrows, but with no articulations. Slight remains of a fleshy cortical
integument.

Three species—in the Coal formation.

†† Leaves only known.

Genus 52. *Poacites.* All Monocotyledonous leaves, the veins of which are parallel, simple, of equal thickness, and not connected by transverse bars.

Several species—in the Coal formation.

Genus 53. *Phyllites.* (Potamophyllites *Ad. Br.*) All Monocotyledonous leaves, the veins of which are confluent at the base and apex, and connected by transverse bars, or secondary veins.

One species—in the Lower freshwater formation.

Obs. M. Ad. Brongniart now refers this fossil to Fluviales; but as it agrees as well with species of Alismaceæ and Butomeæ, we prefer placing it here, under the name originally given it.

††† Fruits only known.

Genus 54. *Trigonocarpum.* Ad. Br.

Five species—in the coal formation.

Genus 55. *Amomocarpum.* Ad. Br.

One species—in the Tertiary formations.

Genus 56. *Musocarpum.* Ad. Br.

Two species—in the coal formation.

Genus 57. *Pandanocarpum.* Ad. Br.

One species—in the Tertiary strata.

———

FLOWERING PLANTS WHICH CANNOT BE WITH CERTAINTY REFERRED TO EITHER THE MONOCOTYLEDONOUS, OR DICOTYLEDONOUS CLASSES.

Genus 58. *Æthophyllum.* Stem simple. Leaves alternate, linear, ribless, not sheathing, having at the base two smaller linear leaflets. (Stipules?) Inflorescence spiked; spikes ovate. Flowers numerous, with a sub-cylindrical tube, or inferior ovarium, and a bilabiate? perianthium with subulate segments.

One species—in the New red sandstone.

Obs. M. Brongniart refers this to Monocotyledons; but if its characters have been rightly determined, it can scarcely belong to that Natural Class.

Genus 59. *Echinostachys.* Inflorescence an oblong spike, beset on all sides with sessile, contiguous, sub-conical flowers, or fruits. *Ad. Br.*

One species—in the New red sandstone.

Obs. M. Brongniart refers this, also, to Monocotyledons, and suggests the possibility of its affinity to Sparganium ; but as there is nothing to show that it is not some Dicotyledonous fruit, such as Datura Stramonium, it will be better to wait for further information before its place is determined on.

Genus 60. *Palæoxyris.* Inflorescence a terminal fusiform spike, with appressed closely imbricated scales; its external portion, where it is not covered by scales, rhomboidal, concave in the middle. *Ad. Br.*

One species—in the New red sandstone.

Obs. One would scarcely think of doubting whether this is Monocotyledonous, so closely does it approach the recent genus Xyris in external characters, if it were not for a tuft of filaments, noticed by M. Brongniart, as apparently proceeding from its apex. This circumstance is at variance with Xyris, and gives rise to a suspicion that it may, perhaps, be some Composita, with a fusiform involucrum.

Class 2. CELLULARES ; or FLOWERLESS PLANTS.

EQUISETACEÆ.

Genus 61. * *Equisetum.* (Oncylogonatum *König.*) Stems articulated, surrounded by cylindrical sheaths, which are regularly tooth-letted, and pressed close to the stem. *Ad. Br.*

One species—in the London clay formation.
One species—in the Variegated marles of the Lias.
One species—in the lower Oolite and Lias.
Two species—in the Coal formation.

? Genus 62. *Calamites.* Stems jointed, regularly and closely furrowed, hollow, divided internally at the articulations by a transverse diaphragm, covered with a thick cortical integument. (? Leaves verticillate, very narrow, numerous, and simple.)

Two species—in the Transition beds.
Several species—in the Coal formation.
Two species—in the New red sandstone.
Two species—in the New red sandstone, and the Coal formations.

FILICES.

Genus 63. *Pachypteris.* Leaves pinnated, or bipinnated; leaflets entire, coriaceous, ribless, or one-ribbed, contracted at the base, but not adherent to the midrib. *Ad. Br.*

Two species—in the inferior beds of the Oolitic formation.

Genus 64. *Sphenopteris.* Leaves bi-tripinnatifid; leaflets contracted at the base, not adherent to the rachis, lobed; the lower lobes largest, diverging, somewhat palmate; veins bipinnate, radiating as it were from the base. *Ad. Br.*

One species—in the Sand below the chalk.
Two species—in the New red sandstone.
Five species—in the Oolitic formation.
Twenty-eight species—in the Coal formation.

Genus 65. *Cyclopteris.* Leaves simple, entire, somewhat orbicular; veins numerous, radiating from the base, dichotomous, equal; midrib wanting. *Ad. Br.*

Four species—in the Coal formation.
One species—in the Transition rocks.
One species—in the Oolitic formation.

Genus 66. *Glossopteris.* Leaves simple, entire, somewhat lanceolate, narrowing gradually to the base, with a thick vanishing midrib: veins oblique, curved, equal, frequently dichotomous, or sometimes anastomising and reticulated at the base. *Ad. Br.*

Two species—in the Coal formation.
One species—in the Oolitic formation.
One species—in the Lias.

Genus 67. *Neuropteris.* Leaves bipinnate, or rarely pinnate; leaflets usually somewhat cordate at the base, neither adhering to each other, nor to the rachis, by their whole base, only by the middle portion of it; midrib vanishing at the apex; veins oblique, curved, very fine, dichotomous —— *Fructification;* sori lanceolate, even, (covered with an indusium,) arising from the veins of the apex of the leaflets, and often placed in the bifurcations. *Ad. Br.*

Twenty-four species—in the Coal formation.
Three species—in the New red sandstone.
One species—in the Anthracite of Savoy.
One species—in the Muschelkalk.

Genus 68. *Odontopteris.* Leaves bipinnated ; leaflet, membranous, very thin, adhering by all their base to
the rachis, with no, or almost no midrib ; veins
equal, simple, or forked, very fine, most of them
springing from the rachis. *Ad. Br.*

Five species—in the Coal formation.

Genus 69. *Anomopteris.* Leaves pinnated ; leaflets linear,
entire, somewhat plaited transversely at the
veins, having a midrib ; veins simple, perpendicular, curved. *Fructification* arising from the
veins, uncertain as to form ; perhaps dot-like, and
inserted in the middle of the veins ; or, perhaps,
linear, attached to the whole of a vein, naked (as
in Meniscia) or covered by an indusium, opening inwardly. *Ad. Br.*

One species—in the New red sandstone.

Genus 70. *Tæniopteris.* Leaves simple, entire, with a stiff
thick midrib ; veins perpendicular, simple, or
forked at the base. *Fructification* dot-like.
Ad. Br.

Three species—in the Lias and Oolitic formations.

Genus 71. *Pecopteris.* Leaf once, twice, or thrice pinnate ;
leaflets adhering by their base to the rachis, or
occasionally distinct ; midrib running quite
through the leaflet ; veins almost perpendicular
to the midrib, simple, or once or twice dichotomous. *Ad. Br.*

Sixty species—in the Coal formation.
Ten species—in the Oolitic formation.
Two species—in the Lias.
One species—in the beds above the Chalk.

Genus 72. *Lonchopteris.* Leaf many times pinnatifid ; leaflets
more or less connate at the base, having a mid-
rib ; veins reticulated. *Ad. Br.*

Two species—in the Coal formation.
One species—in the Greensand formation.

Genus 73. *Clathropteris.* Leaf deeply pinnatifid ; leaflets
having a very strong complete midrib ; veins numerous and simple, parallel, almost perpendicular
to the midrib, united by transverse veins, which
form a net-work of square meshes upon the leaf.
Ad. Br.

One species—in the Lias.

Genus 74. *Schizopteris.* Leaf linear, plane, without midrib, finely striated, almost flabelliform, dividing into several lobes, which are linear and dichotomous, or rather irregularly pinnated, and erect; lobes dilated and rounded towards the extremity. *Ad. Br.*

One species—in the Coal formation.

Genus 75. *Filicites.* This comprehends all that are not referable to the preceding genera.

One species—in the New red sandstone.
Two species—in the Variegated marle of the Lias.

Genus 76. *Caulopteris.* Stem cylindrical, closely marked by large, oblong, convex, uneven scars, wider than the tortuous depressed spaces that separate them.

One species—in the Coal formation.
(One species—in the New red sandstone.)

N.B. It has become necessary to form a new name for this genus, in consequence of all the supposed fern-stems figured or described by Count Sternberg, Ad. Brongniart, and others, under the name of Sigillaria, Favolaria, Rhytidolepis, &c., not being such, as we have elsewhere endeavoured to show. These, of which the nature cannot be doubted, probably belong to species included in some of the genera characterized by the structure of the leaves. The species from the New red sandstone belongs, according to M. Adolphe Brongniart, to Anomopteris Mougeotii.

___

LYCOPODIACEÆ.

Genus 77. *Lycopodites.* (Lycopodiolithus and Walchia *Sternb.*) Branches pinnated; leaves inserted all round the stem in two opposite rows, not leaving clean and well-defined scars. *Ad. Br.*

Ten species—in the Coal formation.
One species—in the inferior Oolitic.
One species—in the sandstone of the Lias?
One species—in the marle below the chalk.

Genus 78. *Selaginites.* Stems dichotomous, not presenting regular elevations at the base of the leaves, even near the lower end of the stems. Leaves often persistent, enlarged at their base. *Ad. Br.*

Two species—in the Coal formation.

Genus. 79. *Lepidodendron.* (Sagenaria.) Stems dichotomous, covered near their extremities by simple, linear, or lanceolate leaves, inserted upon rhomboidal areolæ; lower part of the stems leafless; areolæ (longer than broad) marked near their upper part by a minute scar, which is broader than long, and has three angles, of which the two lateral are acute, the lower obtuse; the latter sometimes wanting.

Several species—in the Coal formation.

Genus 80. *Ulodendron.* Stem covered with rhomboidal areolæ, which are broader than long; scars large, few, placed one above the other, circular, composed of broad cuneate scales, radiating from a common centre, and indicating the former presence of organs that were perhaps analogous to the cones of Coniferæ.

Two species—in the Coal measures.

Genus 81. *Lepidophyllum.* Stem unknown. Leaves sessile, simple, entire, lanceolate, or linear, traversed by a single midrib, or by three parallel ribs; no veins. *Ad. Br.*

Five species—in the Coal formation.

Genus 82. *Lepidostrobus.* Cones ovate, or cylindrical, composed of imbricated scales, inserted by a narrow base around a cylindrical woody axis; their points sometimes dilated and recurved in the form of rhomboidal disks. Seed solitary, oblong, not winged, nearly as long as the scales.

Five species—in the Coal formation.

? Genus 83. *Cardiocarpon.* Fruit compressed, lenticular, heart-shaped, or kidney-shaped, terminated by a sharpish point. *Ad. Br.*

Five species—in the coal formation.

MUSCI.

Genus 84. *Muscites.* Stem simple, or branched, filiform, with membranous leaves, having scarcely any midrib, and being sessile, or amplexicaul, imbricated, or somewhat spreading. *Ad. Br.*

Two species—in beds above the chalk.

CHARACEÆ.

Genus 85. * *Chara.* (Gyrogonites *Lamk.*) Fruit oval, or spheroidal, consisting of five valves twisted spirally; a small opening at each extremity. Stems friable, jointed, composed of straight tubes arranged in a cylinder.

Five species—in beds above the chalk.

ALGÆ.

Genus 86. *Confervites.* Filaments simple, or branched, divided by internal partitions. *Ad. Br.*

Two species—in the Chalk-marle.

Genus 87. *Fucoides.* (Algacites *Schloth.*) Frond continuous, never articulated, usually not symmetrical or subcylindrical, simple or oftener branched, naked or more commonly leafy; or membranous, entire, or more or less lobed, with no ribs, or imperfectly marked ones, which branch in an irregular manner, and never anastomose. *Ad. Br.*

Four species—in the Transition rocks.
Seven species—in the Bituminous shale.
Three species—in the Oolitic formation.
Eleven species—in the Chalk.
Eleven species—in the London clay formation.

PLANTS, THE AFFINITY OF WHICH IS ALTOGETHER UNCERTAIN.

Genus 88. *Sigillaria.* (Rhytidolepis, Alveolaria, Favularia, Catenaria, &c. *Sternb.*) Stem conical, deeply furrowed, not jointed. Scars placed between the furrows in rows, not arranged in a distinctly spiral manner, smooth, much narrower than the intervals that separate them.

About forty species—in the Coal formation.

Genus 89. *Volkmannia.* Stem striated, articulated. Leaves collected in approximated dense whorls.

Three species—in the Coal formation.

Obs. These are possibly the leaves of Calamites.

Genus 90. *Carpolithes.*

Under this name are arranged all the fossil fruits to which no other place is assigned.

LIST OF SUBSCRIBERS.

ABBS, Rev. G. C. *Gateshead.*

Adamson, John, Esq. F.S.A. L.S. R.S.L. &c. &c. *New-castle.*

Alder, Mr. Joshua, *Newcastle.*

Allan, Thomas, Esq. F.R.S. L. & E. *Lauriston Castle, Edinburgh.*

Armstrong, Robert, Esq.

Armstrong, Mr. William, *Killingworth, Newcastle.*

Austen, Sir Henry E. 29, *Cavendish Square.*

Barrow, P. Esq.

Bean, W. Esq. *Scarborough.*

Bedford, His Grace the Duke of, K.G. F.S.A. L.S. G.S. and H.S. *Woburn, Bedfordshire.*

Bell, Thomas, Esq. F.R.S. G.S. and L.S. 17, *New Broad Street.*

Bell, Thomas, Esq. *Picton Place, Newcastle.*

Benson, W. Esq. *Bury St. Edmunds.*

Berkley, Mr. John, *Newcastle.*

Bigge, Charles W. Esq. F.G.S. *Linden, Northumberland.*

Bignold, Samuel, Esq. *Norwich.*

Bigsby, J. J. M.D. F.G.S. *East Retford, Nottinghamshire.*

Bold, Robert, Esq. *Edinburgh.*

Bowman, J. E. Esq. F.L.S. *The Court, near Wrexham.*

Boyd, W. Esq. *Newcastle.*

Boyd, Robert, Esq. *Newcastle.*

b

Broadley, John, Esq. F.L.S. and H.S. *President of the Hull Literary and Philosophical Society.*

Broderip, W. J. Esq. F.R.S. G.S. and L.S. 2, *Raymond's Buildings, Gray's Inn.*

Brockett, J. T. Esq. F.S.A. *Newcastle.*

Bryce, J. Esq. Jun. M.A. Mem. Brit. Assoc. M.G. S.D. &c. *Belfast.*

Buckland, Rev. W. D.D. F.R.S. G.S. and L.S. *Professor of Mineralogy and Geology, Oxford.*

Buddle, John, Esq. F.G.S. *Wallsend, Northumberland.*

Bunyan, R. I. Esq. 6, *Crescent, Blackfriars.*

Burnett, Mr. George, Jun. *Newcastle.*

Carr, George, Esq. *Newcastle.*

Charlton, W. H. Esq. *Hesleyside, Northumberland.*

Charnley, Mr. E. *Newcastle.*

Cheek, H. H. Esq. *Edinburgh.*

Clarke, Rev. W. B. A.M. F.G.S. *East Bergholt, Suffolk.*

Cole, Viscount, M.P. F.R.S. F.G.S. &c.

Cole, Robert, Esq. 33, *Red Lion Square.*

Collingwood, H. J. W. Esq. *Lilburn, Northumberland.*

Cohen, D. W. Esq. *Shacklewell.*

Conybeare, Rev. W. D. M.A. F.R.S. and G.S. Instit. Reg. Soc. Paris. Corresp. *Sully, near Cardiff.*

Copeland, ——, Esq. *Edinburgh.*

Crawhall, Thomas, Esq. *Benwell, Newcastle.*

Corbet, Rev. Waters, M.A. *Longmore Hall, Shropshire.*

Culley, M. Esq. F.G.S. *Coupland Castle, Northumberland.*

Davis, John Ford, M.D. F.L.S. *Bath.*

De la Beche, Henry Thomas, Esq. F.R.S. G.S. and L.S.

Dikes, W. Hey, Esq. F.G.S. *Curator of the Museum of the Hull Literary and Philosophical Society.*

Dixon, Dixon, Esq. *Newcastle.*

Dolphin, John, Esq. *Ruffside, Northumberland.*
Drummond, James L. M.D. Professor of Anatomy, &c. *Belfast.*
Dudley, Miss, *King's Wainsford, Staffordshire.*
Durham, The Right Hon. Lord, F.G.S. and H.S. *Lambton Castle, Durham.*

Egerton, Sir Philip de Malpas Grey, Bart. F.G.S. *Oulton Park, Cheshire.*
Egerton, Thomas, Esq. *St. James's Square.*
Ellicombe, Rev. H. T. M.A. F.A.S. *Bitton Vicarage, near Bristol.*
Empson, Mr. Charles, *Newcastle.*
England, Rev. Thomas, 15, *Surrey Square, Kent Road.*

Falla, W. Esq. F.H.S. and L.S. *Gateshead.*
Fenwick, Thomas, Esq. F.G.S. *Dipton, Durham.*
Ferguson, R. Esq. M.P. *of Raith.*
Fitton, W. Henry, M.D. F.R.S. G.S. and L.S. *Highwood Hill, near Hendon.*
Flounders, Miss, *Yarm.*
Forster, Mr. F. *Haydeck Colliery, near Warrington.*
Foster, Mr. John, *Haswell, Durham.*
Fox, George Townshend, Esq. F.G.S. and L.S. *Durham.*
Fryer, J. H. Esq. *Whitley, Northumberland.*

Gisborne, Rev. Thomas, *Durham.*
Goodhall, H. H. Esq. M.R.A.S. F.G.S. 55, *Crutched Friars.*
Graham, Robert, M.D. F.R.S.E. and L.S. *Professor of Botany in the University of Edinburgh.*
Grantham, Richard, Esq. *Limerick.*

Greenough, G. B. Esq. F.R.S. L.S. & H.S. M.R.A.S. President of the G.S. *Park Road, Regent's Park.*

Guillemard, John Lewis, Esq. M.A. F.R.S. G.S. and L.S. M.R.A.S. 27, *Gower Street.*

Harcourt, Rev. C. G.V. F.H.S. *Whitton Tower, Northumberland.*

Headlam, T. E. M.D. *Newcastle.*

Henslow, Rev. J. S. M.A. F.L.S. and G.S. *Professor of Botany in the University of Cambridge.*

Henry, W. M.D. F.R.S. and G.S. *Manchester.*

Hewitson, Henry, Esq. *Seaton Burn, Northumberland.*

Hewitson, Middleton, Esq. *Newcastle.*

Hewitson, W. C. Esq. *Newcastle.*

Hibbert, Samuel, M.D. F.G.S. *Edinburgh.*

Hill, George, Esq. F.G.S. *Kenton, Northumberland,* (2 copies.)

Hodgson, Mr. R. W. *Newcastle.*

Holland, Henry, M.D. F.R.S. G.S. and L.S. 25, *Lower Brook Street.*

Holroyd, Arthur, M.D. *Harley Street.*

Horner, Leonard, Esq. F.R.S. L.S. and G.S. *Bonn.*

Hoyle, Richard, Esq. *Denton Hall, Northumberland.*

Ingham, Robert, Esq. M.P. F.G.S. *Westoe, Durham.*

Ives, Mrs. *Catton, near Norwich.*

Joplin, Thomas, Esq.

Jukes, Frederick, Esq.

Longlands, J. C. Esq. *Old Bewick, Northumberland.*

Leighton, W. A. Esq. *of Leighton Ville, Shrewsbury.*

Liddell, Hon. Mrs. *Eslington House, Northumberland.*

Lindsay, Dr. James.

Lloyd, George, M.D. *Neachills, Shropshire.*

Lonsdale, W. Esq. F.G.S. *Curator of the Museum of the Geological Society, Somerset House.*

Loch, James, Esq. M.P. F.G.S. 24, *Hart Street, Bloomsbury.*

Losh, James, Esq. *Jesmond, Newcastle.*

Losh, William, Esq. *Benton, Northumberland.*

Mackenzie, Sir G. S. Bart. F.R.S.E. H.S. *Coul, near Dingwell, Rosshire.*

Masters, Mr. W. Jun. F.H.S. *Curator of the Canterbury Museum.*

Monck, Sir C. M. L. Bart. F.H.S. *Belsay Castle, Northumberland.*

Mosley, Sir Oswald, Bart. M.P. F.H.S. *Rolleston Hall, Staffordshire.*

Murchison, R. J. Esq. Vice-President of the Geological Society, F.R.S. and L.S. 3, *Bryanstone Place, Bryanstone Square.*

Murray, Dr. P. *Scarborough.*

Needham, John M. Esq. *Beeston, Nottinghamshire.*

Neill, P. Esq. F.R.S.E. S.A. L.S. G.S. and H.S.M.W.S. *Edinburgh.*

Newby, Rev. Mark, *Witton le Wear, Durham.*

Nicol, Dr. J. J. *Inverness.*

Northampton, The Marquis of, F.G.S. *Castle Ashby, Northamptonshire.*

Northumberland, The Duke of, K.G. F.R.S. S.A. G.S. L.S. and H.S. *Alnwick Castle, Northumberland,* (2 Copies.)

Oakes, James, Esq. F.G.S. *Riddings, Alfreton, Derby.*

Ormerod, G. W. Esq. B.A.
Ormston, Robert, Esq. Jun. *Newcastle.*

Parker, John Cowham, Esq. F.H.S. *Hull.*
Pattinson, H. L. Esq. *Ryton, Durham.*
Phillips, John, Esq. F.G.S. *York.*
Portlock, Captain J. E. R.E. F.G.S. *Depôt Ordnance Survey of Ireland, Dublin.*
Pratt, S. P. Esq. F.G.S. L.S. *Lansdown Place West, Bath.*

Rawson, Christopher, Esq. *Hope House, near Halifax.*
Reddie, John, Esq.
Rennie, Mr. Robert, *Sunderland.*
Rippon, Cuthbert, Esq. M.P. *Stanhope Castle, Durham.*

Salvin, William T. Esq. F.H.S. *Croxdale Hall, Durham.*
Scott, Rev. T. Hobbs, Archdeacon of New South Wales, F.G.S. *Whitfield Rectory, Northumberland.*
Sebright, Sir J. S. Bart. M.P. F.G.S. and H.S. *Beechwood, near Market Street, Herts.*
Sedgwick, Rev. A. M.A. F.R.S. G.S. *Fellow of Trinity College, and Woodwardian Professor of the University of Cambridge.*
Sharpe, D. Esq. F.L.S. G.S. *Lisbon.*
Silvertop, George, Esq. F.H.S. *Minsteracres, Northumberland.*
Steggall, Rev. W. A.M. *Bury St. Edmunds.*
Stephenson, George, Esq. *Alton Grange, near Ashby de la Zouch.*
Stephenson, Robert, Esq. *Newcastle.*
Stirling, W. F. Esq. 5, *New Square, Lincoln's Inn.*
Stobart, Henry, Esq. *Pelaw, Durham.*

ix
Stokes, C. Esq. F.R.S. S.A. G.S. and L.S. M.R.A.S.
4, *Verulam Buildings, Gray's Inn.*
Straker, John, Esq. *Jarrow Lodge, Durham.*
Surtees, Anthony, Esq. *Hamsterley Hall, Durham.*
Swinburne, Sir J. E. Bart. F.R.S. S.A. *Capheaton, Northumberland.*
Swinburne, Lady, *Capheaton, Northumberland,* (2 Copies.)
Swinburne, Thomas, Esq. *Gateshead.*

Taylor, William, Esq.
Thorp, Rev. Charles, Archdeacon of Durham, *Ryton.*
Torrie, T. J. Esq. *Edinburgh.*
Trollope, Henry, Esq. *Harrow.*
Turner, Rev. W. M.A. F.G.S. *Trinity College, Cambridge.*
Turner, Rev. W. F.G.S. *Newcastle.*
Turner, J. A. Esq. *Manchester.*
Trevelyan, W. C. Esq. M.A. F.L.S. and G.S. *Wallington, Northumberland.*

Vigors, N. A. Esq. M.P. M.A. Sec. Z.S. F.R.S. S.A. G.S. L.S. H.S. and M.R.I.A. *Regent's Park.*

Warburton, H. Esq. M.P. M.A. Vice-President G.S. F.R.S. & H.S. 45, *Cadogan Place, Sloane Street.*
White, H. C. Esq. F.G.S.
Williams, Rev. D. *Bleadon, near Cross, Somersetshire.*
Williamson, J. W. Esq. *Whickham, Durham.*
Winch, N. J. Esq. A.L.S. F.G.S. *Newcastle.*
Witham, H. T. M. Esq. F.R.S.E. G.S. &c. *Lartington Hall, Yorkshire.*
Wood, George W. Esq. M.P. F.G.S. *Manchester.*

Wood, Nicholas, Esq. *Killingworth, Newcastle.*
Wood, Thomas, Esq. *Hetton, Durham.*

Youens, Rev. Dr. *Ushaw College, Durham.*

Belfast Natural History Society.
Geological Society of London.
Literary and Philosophical Society, Newcastle.
Literary, Scientific, and Mechanical Institution, Newcastle.
Literary and Philosophical Society, Leeds.
Nottingham Subscription Library.
Ratcliffe Library, Oxford.
Scarborough Philosophical Society.
University of London.
York Subscription Library.
Yorkshire Philosophical Society.

PINITES BRANDLINGI.

THE WIDEOPEN FOSSIL TREE.

––––––

Witham, *Observations upon Fossil Vegetables, p.* 31. *tab.* 4. *figs.* 1, 2, 3, 4.

––––––

This plate represents a portion of an immense fossil, which has lately occurred in a grindstone quarry, at Wideopen, near Gosforth, about five miles north of Newcastle-upon-Tyne. The bed in which this quarry is worked, is considered one of the highest members of the coal formation; and has its name of " Grindstone Bed" from being extensively quarried for millstones and grindstones. The fossil measured, in its whole length, seventy-two feet; the portion figured being of the lower end, and not quite one half the length. It followed a bed or parting in the stone, and lay nearly at right angles with the dip of the strata, which is a little to the southward of east; its direction was nearly north-east and

south-west, the lower end being towards the
north-east; it tapered gradually from the bottom,
which was four feet nine inches, to the top,
which was eighteen inches in breadth. The substance of the fossil, where the wood was perfectly
petrified, was of a silicious nature, and of a brown
colour, having well defined crystals of ferruginous
quartz, in cavities interspersed through it, and
differing entirely from the surrounding mechanical
deposit of sandstone. A few fine veins of white
quartz, approaching chalcedony, passed through
the substance of the fossil longitudinally. A
thickness of thirty feet of solid stone had been
worked away before it was discovered, which was
at first accidentally, in the operations of the
quarry. It is to the scientific zeal and liberality
of the Rev. R. H. Brandling, on whose estate the
quarry is, that we are indebted for our complete
knowledge of this fossil giant of the vegetable
kingdom; he having, at considerable trouble and
expense, caused it to be laid open to its full extent,
with the greatest care; and, besides, having had
an artist upon the spot, who took a drawing before
any attempt was made to move it from its bed.

When that portion of the stone which covered
the fossil was carefully removed, there appeared
a dull black carbonaceous substance, soft and
wet, and which soiled the fingers; this completely
enveloped the whole fossil, and was a little more
than an inch thick, but without any markings

upon the surface indicative of the bark or outside of the plant. Beneath this was a layer of a bright ochrey yellow substance, also soft and wet, but which, when undisturbed, shewed marks of the woody fibre ; these two substances coated the whole of the fossil, but never intermixed. Beneath the yellow matter was the petrified wood ; and when the double coating was removed, so complete was the likeness of the fossil to the trunk of a decayed tree, that any one would have been tempted to try it with a knife. The relative situation of the different parts just described, will be best understood by a sketch, where

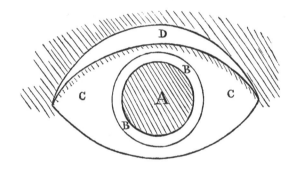

A is the petrified wood.
B the yellow substance.
C the black ditto.

It must be observed, that the cavity in which the fossil lay was never circular, but had one dia-meter longer than the other, apparently from compression, which generally had caused it to assume the pointed shape shewn in the sketch, these points being completely filled with the black matter; there was, also, frequently an empty space between the upper side of the fossil and the covering of stone, as if from the shrinking or diminution of bulk in the tree, as shewn in the sketch, at **D**.

It is difficult to account for the strongly marked difference of colour in the two enveloping substances, unless we suppose the black to have been the bark, and the yellow to have arisen from the decay of part of the woody fibre, before the slow petrifying process, by which the silicious particles were substituted for those of the wood, had time to operate. It is probable the outer part all round, as well as portions of the whole tree, had been in a state of decay before it was deposited where we now find it, as the whole of the fossil, for six feet in length at the lower end, was composed of the soft, black, and yellow substances, just described; the black always forming the outer coating only, and the yellow being substituted for the entire woody part. Where this occurred, the compression was very great, the breadth being at the lowest part four feet nine inches, whilst the perpendicular thickness was only nine inches; the higher end of the fossil was also entirely composed of these substances, and very much flattened; in one or two places in the length, also, the woody fibre was almost entirely changed into them. No roots could be seen, nor any thing like branches, except the large knots shewn in the drawing. The greatest diameter of the petrified part was about two feet. In attempting to move it, notwithstanding every possible care and anxiety to preserve it whole, it fell to pieces, so that the largest fragment obtained is not above eighteen inches long, and displays little more than half the diameter of the fossil.

Many impressions of Calamites occur in the sandstone of this quarry ; and it is worthy of observation, that their whole substance is composed of the same kind of black powdery carbonaceous matter which covered the outside. A thin seam of coal occurs immediately under the sandstone, and about seven feet below the bed of the fossil.

There is a striking difference between the nature of this fossil, and of those we usually find in the sandstone or shale beds of our coal fields, where we generally have a cast of the outside form only, without the least indication of internal organization, their substance being of the same nature as that of the rock in which they occur; but here the case is entirely different. The outer form is ill defined, whilst the internal structure, even to the minutest vessel, is perfectly preserved. This circumstance sufficiently indicates the difference of the nature of the two classes of plants ; one was of a soft membranous texture, easily yielding to pressure and decay, whilst the strong woody fibre of the other would long withstand both, and allow that gradual substitution of matter, by infiltration, by which alone delicate internal organic structures can be preserved.

Explanation of Plate 1.

a and *b,* the lower portion of the fossil, as it appeared when the covering of stone and the powdery enveloping substances were removed.

b to *c,* a continuation of the bed of the fossil.

c, a section of the lower half of the bed of the fossil.

f, part of the lower portion of the tree on an enlarged scale.

g, section of ditto.

Our tree was plainly, judging from its external appearance alone, Exogenous; of this the irregular arrangement of the knots, indicating the origin of branches upon its trunk, and its manifest tendency to a conical form, are sufficient evidence. This is confirmed by an examination of its anatomical structure, which is nearly as perfect as in recent wood.

Beautiful figures of the appearance of its tissue in a transverse section, have been given by

Mr. Witham; and, in this view, it is so extremely similar to that of Coniferæ*, that it might almost be inferred that the tree actually belonged to that tribe. But, in the first place, neither Mr. Witham, nor ourselves, have been able to discover any trace of concentric circles; and, secondly, a longitudinal section, in the direction of its medullary rays, shews that the woody fibre (or elongated cellular tissue) of the trunk differs in some important particulars.

In Coniferæ, the walls of the woody fibre are occupied by a peculiar kind of pore-like glands, by which they are distinguished from all other tribes of recent plants, except Cycadeæ;† these glands may be readily seen by inspecting a thin shaving of pine wood, or reference may be made to excellent figures of them in Kieser's *Memoire sur l'organisation des plantes, tab.* 15, *fig.* 74, *b* and *c*, &c. But in the tissue of this fossil, no such structure is discoverable. On the contrary, the walls of the woody fibre are beautifully reticulated, or covered with hexagonal meshes, a structure with nothing analogous to which are we acquainted in the wood of recent plants. This is represented at fig. 2, where a small portion of the tissue is very highly magnified.

The anatomical structure of this fossil is more

* *Introduction to the Natural System of Botany, p.* 247.

† *Introduction to the Natural System of Botany, p.* 245.

perfectly preserved than of that which forms the subject of the next plate ; but they are, nevertheless, so extremely similar, that no doubt can exist of their both being, if not the same species, at least very nearly allied. The principal difference between them, consists in the reticulations of the woody fibre of this fossil being more regular and larger than those in the Craigleith plant.

PINITES WITHAMI.

CRAIGLEITH FOSSIL TREE.

WITHAM, *in the Philosophical Magazine and Annals, for January*, 1830. THE SAME, *Observations on Fossil Vegetables, p.* 30. *tab.* 3. *figs.* 8, 9, 10, 11, 12.

Found in the year 1826, in the great quarry, at Craigleith, near Edinburgh, which we take to be in a sandstone, considerably below the coal formation proper; perhaps, even, in the mountain limestone group. It was thirty-six feet long, and three feet in diameter at its base; its position was nearly horizontal, or corresponding with the dip. No branches were discovered, although there are indications of them upon that portion of the fossil which yet remains. Unlike the Wideopen fossil, last described, the outward form of this was entire, the bark being converted into coal. The mineralizing substance was principally carbonate of lime, the crystallization of which had broken up and distorted the fine vegetable tissue in a considerable portion of the fossil.

The figures represent highly magnified views of portions of the wood, drawn from specimens prepared by Mr. Sanderson, of Edinburgh.

1. Exhibits a longitudinal section, made in the direction of the medullary rays, some of which are seen adhering to it, and lying across it. The tissue consists of elongated cellules (woody fibre) fitted together by their rather abruptly pointed extremities, and very like those of Cupressus sempervirens. There is no trace of any kind of vessel passing through this tissue. The membrane of the cellules has, now and then, a reticulated appearance, as if it had been itself composed of extremely minute spheroidal cellules, or filled with such; these are, occasionally, apparent, where the specimen is sufficiently transparent to allow light to be transmitted freely, except when the membrane of the tissue has been destroyed in the grinding down and polishing; they seem to have been smaller than in the Wideopen fossil, but are by no means so beautifully preserved. Whether this difference in the reticulation was connected with external characters, we have, at present, no means of judging; there is no trace of any other kind of organization. The medullary rays are merely indicated by various transverse bars, such as are represented.

2. Seems to be a longitudinal tangental section, in which the elongated cellules of the wood

appear more distorted and injured, than in
the last; and the passages of the medullary
rays, from the centre to the circumference, are
distinctly cut through. These passages are
variable in size, sometimes appearing to consist
of as many as four layers of muriform cellules placed side by side, sometimes not having
more than two. The same reticulated structure
of the membranes of the elongated cellules of the
wood, as described in figure 1, is more or less
visible in places; and would, no doubt, have been
equally so, if the specimen described had not been
ground down so much thinner. The mouths of
the medullary passages are from the 200th to the
400th of an inch across.

3. Shews the appearance of a minute portion of
a transverse section, highly magnified, with three
medullary rays, one of which consisted, at the
line where it was cut through, of four layers of
muriform cellules. The mouths of the elongated
cellules of the wood are unequal in size, but
average the 400th of an inch in diameter. No
trace was discoverable either of concentric zones,
or of the orifices of ducts.

4. Is an ideal figure, to explain the parts whence
these sections are supposed to have been taken.
a, refers to fig. 2.   *b*, to fig. 1.   and *c*, to fig. 3.

Our observations on the specimens we have
seen of this fossil, agree entirely with those of

Mr. Witham, in his beautiful work on Fossil Vegetables, above quoted; especially in the absence
of any trace of concentric zones in the wood. In
a polished piece, four inches and a half across,
nothing of the kind is to be detected. Mr. Witham
correctly observes, that, in every thing else, in a
horizontal view, the accordance between this plant
and Coniferæ is perfect; but in a longitudinal
section of Coniferæ, the walls of the woody tissue
are, as has been before stated, in all the recent
species that have been examined, distinctly
marked with circular elevations, equal to about
half the breadth of the elongated cellules, each
having the appearance of a perforation in its
centre. Of these circular elevations, no trace is
discoverable in this fossil; on the contrary, the
walls appear, as above described, to consist of
very minute cellules, arranged in a reticulated
manner. We are, therefore, compelled to conclude,
notwithstanding the great similarity between the
transverse sections of this wood, and those of recent
Coniferæ, and notwithstanding the total absence of
ducts, in what seems to have been a tree, having
an Exogenous structure, yet that as the very remarkable organization of the walls of the woody
tissue of recent Coniferæ does not exist in this
fossil, but is supplied by another kind of structure
of an equally unusual nature, the inference that
this tree belonged to the Coniferous tribe, cannot
be considered altogether just.

PINITES MEDULLARIS.

A CRAIGLEITH FOSSIL BRANCH.

WITHAM, *in the Transactions of the Natural History Society
of Northumberland, Durham, and Newcastle-upon-Tyne.
Vol.* 1. *p.* 297. *tab.* 25. *figs.* 3, 4, 5, 6, 7, 8.

This represents the tissue of a branch found
in the same quarry, at Craigleith, as the last,
in the early part of the present year, 1831. A
fragment of a stem occurred in one of those hard
indurated masses, not uncommon in this quarry,
which are very difficult to work; in this instance,
powder was used, which probably detached the
branch, part of which is here figured; but this
could not be satisfactorily ascertained.

In this specimen, the concentric circles, medullary rays, and pith of an Exogenous tree, are
distinctly seen; otherwise, the appearance of the
tissue is as nearly as possible that of the tree from
the same quarry, figured at plate 2. It is, how-

ever, remarkable, that in a specimen no more than half an inch thick, there are distinct traces of four concentric zones, while, in the plant represented at plate 2, there is not an appearance of a single zone in a specimen four inches and a half thick.

The principal difference between this and the usual structure of Coniferous wood, is the large proportion that is borne to the zones by the pith, which is four times greater in diameter than the first zone of wood that surrounds it. We are not acquainted with any recent Coniferous species in which so great a difference between the woody zones and the pith has been observed. We have seen no longitudinal section of this specimen; it is, therefore, uncertain whether the tissue agrees in other respects with that of the large stem of the Craigleith quarry, figured at plate 2.

Fig. 1, is a view of the section of the branch of the natural size, as it appears to the naked eye, with the pith and the concentric circles.

Fig. 2, is a portion of the same, very highly magnified, shewing the structure of the pith, the medullary rays, and the mouths of the cellules of the woody tissue. In this, the medullary sheath is so much converted to a coaly matter, that its exact structure is no longer to be detected.

LEPIDODENDRON STERNBERGII.

Lepidodendron dichotomum. *Sternberg essai d'un exposé geognostico-botanique, p.* 25. *tab.* 1. *and part of tab.* 2.

L. Sternbergii. *Ad. Brongniart Prodrome d'une histoire des Végétaux Fossiles, p.* 85.

The specimen here figured is from the shale, forming the roof of the low main coal seam, in Felling Colliery, near Newcastle-upon-Tyne. Vegetable fossils occur in all the sandstone and shale beds of the coal formation, and in many of the members of the subjacent mountain limestone group. The coal itself very rarely retains any marks of organic structure. In many of the sandstones, although the fossils are numerous, it is only the large and strongly marked individuals which have left their forms impressed upon the rough-grained mechanical deposit of these rocks, when their bark or outer coating is generally found converted into a fine coal.

The limestone itself has, hitherto, afforded but few vegetable remains; nevertheless, we shall have to notice, in the progress of this work, some beautiful examples, both from the limestones of Northumberland, and from those in the neighbourhood of Edinburgh, which are curious from their intimate connexion with the remains of animals. It is the beds of shale, or argillaceous schistus, which afford the most abundant supply of these curious relics of a former world ; the fine particles of which they are composed, having sealed up and retained, in wonderful perfection and beauty, the most delicate outward forms of the vegetable organic structure.

Where shale forms the roof of the workable seams of coal, as it generally does, we have the most abundant display of fossils; and this, not, perhaps, arising so much from any peculiarity in these beds, as from their being more extensively known and examined than any others. The principal deposit is not in immediate contact with the coal, but about twelve to twenty inches above it; and such is the immense profusion in this situation, that they are not unfrequently the cause of very serious accidents, by breaking the adhesion of the shale bed, and causing it to separate and fall, when, by the operations of the miner, the coal which supported it is removed. After an extensive fall of this kind has taken place, it is a curious sight to see the roof of the mine covered with these vegetable forms, some of them of great beauty and delicacy; and the observer cannot fail to be struck with the extraordinary confusion, and the numerous marks of strong mechanical action exhibited by their broken and disjointed remains. The Lepidodendrons are, after the Calamites, the most abundant class of fossils occurring in the coal formation of the North of England, and are sometimes of a large size, fragments of stems occurring from 20 to 45 feet long; we have, ourselves, measured in the shale forming the roof of the Bensham coal seam in Jarrow Colliery, an individual of this class, four feet and a half in breadth.

In examining the species of Lepidodendron, a botanist finds four characters by which he may compare them with recent plants, viz., their *surface*, their *foliage*, their *ramifications*, and their *texture*.

It is with Coniferæ, and Lycopodiaceæ,* that Lepidodendrons have to be compared in all these particulars.

With regard to their surface, in both Coniferæ and Lycopodiaceæ, the leaves have a similar arrangement, and the scars, or marks, caused by the fall of the latter, are of a similar kind. In Coniferæ, the leaves are arranged upon the stem, in two very different ways. First, in the species

---

* *Introduction to the Natural System of Botany, p.* 316.

having, what botanists denominate, fascicled foliage, such as the Scotch Fir (Pinus Sylvestris), the Pinaster (Pinus Pinaster), the Weymouth Pine (P. Strobus), and the like, the first leaves that are developed are brown and membranous, roll back and wither away, almost immediately after the young branch has acquired its first growth. From the axilla of each of these, sprouts forth a bud, that never or rarely elongates, but which produces several leaves, the outermost of which are membranous and perishable like the first; but the innermost, narrow and rigid, forming the permanent green foliage of the species; in these, where the foliage has fallen away, the stem is covered with numerous narrow projections, thickest at the upper end, where the remains of withered leaves are visible, arranged, spirally, with great symmetry, and separated by intervals, usually equal, at least to twice the breadth of the projections.

Secondly, In the species in which the leaves are solitary, as in the Spruce fir, the Araucaria, the Cunninghamia, &c., the leaves that are originally developed when the young shoot forms, never undergo any material alteration, but are those which subsequently become the green foliage of the plant; none, or few, apparent axillary buds are developed ; and, finally, the leaves either separate by a clean scar of a rhomboidal or roundish figure, with a depressed point in its middle, where the vascular bundle connecting the stem and leaf was broken through, or separate imperfectly, leaving behind an irregular mark upon a rhomboidal areola. The yew is an instance of the former; Cunninghamia and Araucaria of the latter. In all cases, the scars, or the rhomboidal areolæ, are disposed in a spiral manner, with the most exact symmetry. With Coniferous plants of the latter kind, Lycopodiaceæ accord so much in the arrangement of their leaves, and, consequently, in the appearance of the surface of the stems, after the leaves have fallen, that it would be difficult to point out any difference, except that they are often, as in Lycopodium clavatum, rigidum, divaricatum, &c. less spiral, having a tendency to become verticillate. Lepidodendra accord equally with Coniferæ, and Lycopodiaceæ, in the arrangement of the scars of the leaves.

The foliage of certain Coniferæ, such as Araucaria, and of Lycopodiaceæ, is so similar, that their casts would be scarcely distinguishable, except by the larger size of the former. Lepidodendra accord better with Coniferæ than with Lycopodiaceæ, in this respect.

The ramifications of Coniferæ and Lycopodiaceæ are essentially different. In the former, the branches arise from the same plane on opposite sides of the main stem, often assuming a verticillate arrangement. In the latter, the branches bifurcate, whenever a new bud is brought into action, so that the whole of the divisions are dichotomous; and the same takes place in the inflorescence whenever the latter is compound, as in L. Phlegmaria. Hence, Lepidodendra are more related to Lycopodiaceæ than to Coniferæ, in their manner of branching; and as dichotomous ramifications are extremely rare in recent plants, this circumstance, taken together with their other characters, strengthens M. A. Brongniart's opinion of their strong analogy with Lycopodiaceæ.

The texture and size of Lycopodiaceæ and Coniferæ are very dissimilar. The former are soft cellular plants, with small creeping or erect stems, no bark, and an imperfect formation of a woody axis; the latter are large trees, with a thick bark, and a hard woody centre, which is incapable of compression by any ordinary force. With neither tribe do Lepidodendra agree in these points: they resemble Lycopodiaceæ in their soft stem; for specimens, some inches in diameter, are found so compressed, as to be nothing more than a thin plate; but they agree with Coniferæ in the size they seem to have attained, and in the presence of bark, although that part is thin, compared with the bark of recent Coniferæ.

Upon the whole, we are led to conclude, that the Lepidodendron genus was not exactly like either Coniferæ or Lycopodiaceæ, but that it occupied an intermediate station between those two orders, approaching more nearly to the latter than to the former.

The species now represented appears not to be distinct from the L. Sternbergii of A. Brongniart; the broader figure of the areolations of the specimen, represented by De Sternberg, being, probably, due to their younger state.

The rhomboidal spaces were doubtless the base of the leaves, which appear to have been linear lanceolate, and slightly curved. The depression a little above the middle of the spaces, was where these leaves separated; and the obscure line, that runs from the depression towards, or to the base of the rhomboidal space, was a depression originally, and does not seem to have had any relation to the organic union of some part now obliterated. The depression, in some specimens, is evidently caused by the breaking off of the fossil leaves from the axis, where the bed, in which the specimen laid, was divided.

ULODENDRON MAJUS.

RHODE *Beiträge zur Pflanzenkunde der Vorwelt, t.* 3. *f.* 1.

The specimen here figured, is from the shale forming the roof of the Bensham coal seam, in Jarrow Colliery, near Newcastle-upon-Tyne.

It would be probable that this was an old stem of a Lepidodendron, and even, perhaps, of L. Sternbergii, notwithstanding the areolæ of the surface being different in figure from those of that species, if the figure of the areolations were altered by age in Lepidodendron, as in recent Coniferæ. In the latter, areolæ, which, when young, have their perpendicular diameter the greatest, alter into rhomboids, having their horizontal diameter the longest ; a circumstance which arises from the tissue of the bark being strained horizontally by the formation of new wood beneath it. But M. Adolphe Brongniart has well observed, (*Prodrome*, p. 84,) that the scars of Lepidodendron, instead of shortening, lengthen during the growth of the plant, as is shown in the first plate of Count Sternberg's Essai, and as is also seen in plates 4 and 9 of this work, whence he infers, that Lepidodendra did not increase in diameter. Without adopting this conclusion, we have no difficulty in recognizing the accuracy of the observation. It is therefore probable, that this fossil, although very similar to a Lepidodendron, was really of a different nature. At all events, it contains evidence of its genus having been very unlike any thing we have among Coniferæ, or Lycopodiaceæ; for in these two orders we have nothing that can be compared to those large scars upon the surface of this specimen, which indicate points whence branches, or, more probably, masses of inflorescence must have fallen. It would seem that these lost portions, whatever they were, consisted of scales, imbricated closely over each other around a common woody axis, in the same way as the scales of the cone of a Pinus; and it also appears that the scales had a figure different from the leaves. There are connected with these scars two considerations of much importance; viz. 1. That the supposed masses of inflorescence were not only neither terminal, nor disposed spirally upon the stem, but were also produced upon the old trunks, and not upon the young branches; circumstances at variance with any thing we know of recent Coniferæ, or Lycopodiaceæ; and 2. That the scars are placed one beneath the other, and not spirally, or alternately, upon the stem. The furrows upon the surface of the specimen shew, that it has been pressed from a cylindrical into a flat figure.

For the convenience of speaking of fossils of this kind, we have provisionally called them by a name suggested by their scars, notwithstanding a possibility of their being old stems of Lepidodendron ; and we have done this with the less reluctance, seeing that the nomenclature of fossil botany must, for some time, be necessarily merely provisional to a great extent.

6

ULODENDRON MINUS.

ALLAN *in Edinburgh Philosoph. Trans. vol.* 9. *p.* 235. *t.* 14.
Lepidodendron ornatissimum. *Ad. Brong. prodr. p.* 85.

The specimen from which this figure is taken is
in shale, from the roof of the high main coal seam
at South Shields Colliery, county of Durham.
That figured by Mr. Allan, in the Edinburgh
Philosophical Transactions, was from the Craigleith quarry.

It is most likely a younger state of Ulodendron majus ; we, nevertheless, distinguish it,
because, in recent plants, the size of the masses
of inflorescence, in the same species, is not materially different, whatever may be the age of the
individual that bears them; while, in this, the
scars indicate traces of lost bodies, that are
scarcely half as large as those of Ulodendron
majus. The stem has been pressed flat, and both
sides of it are preserved in the specimen figured;

on the opposite side to that which has been drawn is a similar row of scars, having the same arrangement. It would be probable, that this specimen is represented in an inverted position, if we were sure that the laws of its structure were the same as those of recent plants; but there is no satisfactory evidence upon this point.

Fɪɢ. 1.

LEPIDODENDRON ACEROSUM.

———

See Plate 8.

From shale, in the roof of the Bensham Colliery.

Fɪɢ. 2.

LEPIDODENDRON DILATATUM.

From the roof of the low main coal seam in Felling Colliery, near Newcastle-upon-Tyne.

This fossil is scarcely referable to any figured Lepidodendron. It appears to have been the fragment of the apex of some dilated species, which has been compressed, without the arrangement of its parts having been disturbed. Possibly, it may be the same as that represented by the upper left hand figure of the second plate in Count Sternberg's *Flore du monde primitif*. The leaves are longer than in L. Sternbergii, to which that author considered his to belong.

FIG. 3 and 4.

LEPIDOPHYLLUM LANCEOLATUM.

Both these specimens are from the shale, forming the roof of the Bensham coal seam at Jarrow Colliery.

It is probable that these leaves belonged to some species of Lepidodendron; they were evidently of a woody rigid texture, had a middle rib, and were triangular at the base, becoming flat upwards. Perhaps, it is to Lepidodendron? acerosum, that they should be referred.

LEPIDODENDRON ? ACEROSUM.

The species here figured, is from the shale, forming the roof of the low main coal seam at Felling Colliery, near Newcastle-upon-Tyne.

Both these specimens, and that represented at figure 1, of the last plate, are, no doubt, fragments of the same species. We refer them to Lepidodendron with doubt, because of the absence of any proof of their stems having been dichotomous, and on account of the irregular manner in which the leaves originated. We do not find that uniformity of size in the scars, and that symmetrical arrangement, which are so characteristic of true Lepidodendra; but instead, an imperfectly spiral appearance, which does not seem to be owing to the specimen having been crushed, but to have been part of its original nature. This is particularly apparent in figure 1; and also, in figure 1, plate 7.

LEPIDODENDRON GRACILE.

———-

From shale, in the roof of the low main coal seam, Felling Colliery. A fine specimen is in the possession of the Geological Society.

This beautiful fossil gives a good idea of what a true Lepidodendron was, and exhibits a more distinct approach to Lycopodium, especially such species as L. squarrosum, than the larger species. It resembles Lepidodendron Sternbergii in many respects, but seems to have been more slender, and to have had smaller leaves, leaving more acutely rhomboidal scars.

Fig. 2 represents a portion of one of the branches, of the natural size.

LEPIDOSTROBUS VARIABILIS.

All the specimens here figured, are from the shale, forming the roof of the Bensham coal seam, at Jarrow Colliery.

These are, no doubt, bodies analogous to that figured by Parkinson, in his organic remains, (vol. 1. pl. 9. f. 1.) which M. A. Brongniart calls Lepidostrobus ornatus, and which he considers as a cone, the scales of which are terminated by rhomboidal disks, imbricated from above downwards.

The arrangement of the terminations of the scales in this species is certainly the reverse; the scales are sharp pointed, and are imbricated in the usual way, all their ends turning towards the apex of the cone.

It might be supposed, that it was such bodies as these that left behind them the scars on the Ulodendra, tab. 5 and 6, and that they were really the fructification of that genus. Upon this subject we shall have some observations to offer, in discussing the structure of the fossil, represented at tab. 11.

11

LEPIDOSTROBUS variabilis.

Lepidostrobus variabilis. *Suprà, tab.* 10.

From the roof of the Bensham coal seam, at Jarrow Colliery.

These specimens exhibit other states of the fossil represented in the last plate. That they all belong to the same species, we can scarcely doubt, considering their being constantly found together ; and that their differences are apparently dependent only upon their different ages ; thus tab. 10. fig. 2., *a.* and fig. 1. of this plate, are very young ; tab. 10. fig. 2. *b.* is rather older; fig. 3. tab. 10. may be the same thing further developed, the conical termination visible in the young specimens having changed into a rounded one. Fig. 1. tab. 10., shews the fossil at a yet more advanced age ; and we have a specimen now before us still longer, with the end doubled back, in consequence, as it would seem, of some pressure.

A conical axis, around which a great quantity of scales were compactly imbricated from the base upwards, was obviously the structure of this species of Lepidostrobus. These scales were narrow, and gradually acuminated, as represented at tab. 11. fig. 1. when young ; but when older, they appear, from other specimens, to have become broadly ovate, with a rigid mucro. That the scales were imbricate from below upwards, their points being directed to the apex of the cone, is evident, in young specimens, such as fig. 1. tab. 11. and, indeed, from many older specimens also. Sometimes, however, they are apparently turned downwards, a circumstance that is owing to their having been forcibly compressed from above downwards; an instance of this is given at fig. 1. *a.* tab. 10., in which the left side is in such a state, while the right side retains its natural position. Their axis appears, notwithstanding its thickness, to have been soft and pliable ; at least, such an inference seems warranted by the specimen before alluded to, in which the cone is bent almost double, without any fracture, an inch and a half below its apex ; a circumstance which certainly would not have taken place in any part, of which the axis was woody and rigid.

Mr. Adolphe Brongniart entertains no doubt of these cones being reproductive bodies, analogous to those of recent Coniferæ, and Lycopodiaceæ; and, it is probable, that this view of their nature is correct ; at the same time, it must be confessed, that if all the species had, like the original species, scales, with a dilated reflexed, rhomboidal disk, it might be a matter of doubt whether they were not more nearly related to Cycadeæ.* It may further be remarked, that in some specimens, there seems to have been a cylindrical, or oblong body, lying in the axilla of each scale ; and, in such instances, the appearance of the fossil is very like that of a young shoot of the genus Pinus, before the first ramentaceous leaves are pushed aside by the secondary green permanent ones, (see page 18, at the top) ; nor is this the only important point of resemblance between Lepidostrobus, and the young shoots of Pinus ; the latter vary much in appearance, according to their age ; and their ramentaceous scales, which point upwards when young, roll backwards when older. Such shoots are, also, very flexible ; and their axis, when stripped of the scales, has scars, arranged much in the same manner as in the fossil. So striking, indeed, are these analogies, that there is only one point that decides us in adopting M. Brongniart's view, namely, that the Lepidostrobi were always articulated with their stem ; a circumstance which is common in those masses of inflorescence, which Botanists call amenta, or strobili, and to which

* ' *See Introduction to the Natural System of Botany,* p. 245.

Lepidostrobi must be referable, but which is extremely rare in mere branches.

As Lepidostrobi may be considered to have been almost proved to be organs of fructification, it is a point of great moment to discover to what other fossil remains they appertain. In the opinion of M. Brongniart, they undoubtedly belong to Lepidodendron; and supposing that Ulodendra could be shewn to be old stems of Lepidodendra, we should entirely agree with him; for, although no one has succeeded in discovering Lepidostrobi, except in their detached state, yet there is so much resemblance between the base of these cones, and the scars of Ulodendron, that one can hardly doubt their having been separated from each other. We have a specimen of the base of what appears to have been a Lepidostrobus, from the Barnsley coal field, given us by Mr. Edgar, which is so like in size and structure the lower scar of Ulodendron minus, tab. 6., that they actually look as if one had been broken off the other. But even in regard to the identity of Ulodendra and Lepidostrobi, there is this difficulty, that while the latter are very common, the former are extremely rare; and in taking M. Brongniart's view of the question, the difficulty seems increased. In the first place, it has been shewn, (p. 20-21,) that the affinity of Lepidodendra, judging from their stems and leaves only, is greater with Lycopodiaceæ than any thing else that is recent. Now, this opinion is incompatible with the Lepidostrobi belonging to Lepidodendra, because the fructification of Lycopodiaceæ consists in a mere alteration of the leaves at the ends of the branches, without any dis-articulation ever, in any known instance, taking place. Moreover, the fructification of Lycopodiaceæ is always terminal ; and, although we have numerous well preserved ends of Lepidodendron branches, no one has seen them assuming the appearance of a Lepidostrobus. Another difficulty in the way of M. A. Brongniart's supposition, is, that Lepidostrobi are much more common in company with Ferns and Calamites, than with Lepidodendra. Of four large specimens now before us, containing impressions of Lepidostrobi, there is not a trace of a Lepidodendron ; in one specimen, a single large cone lies among fragments of ferns; in a second, we have five Lepidostrobi, with a few indistinct casts, either of a Calamites, like C. arenaceus, or of the stalk of some large fern-leaf; in a third, there are nine Lepidostrobi, with a morsel of some Calamites, and a fragment of the leaf of some Neuropteris ; while, in the fourth, a single cone lies among fragments of Calamites, and various fern stems.

We shall take an early opportunity of returning to this enquiry. In the mean while, we would particularly direct the attention of Geologists to the importance of discovering these bodies actually *attached* to the plants to which they belong. Such is the uncertainty of all these inquiries, that, until the cones shall have been discovered in such a state, any view of the subject must be extremely conjectural.

LEPIDODENDRON SELAGINOIDES.

L. selaginoides. *Sternberg essai d'un exposé géognostico-botanique, &c. p. 35. t. 16. f. 3. and t. 17. f. 1. Ad Brongn. prodrome, p. 85.*

" Pinus sylvestris Mugo Tabernæmontani et
Mathioli. *Volkm. Siles. subterr. t. 12. f. 6.*"
" Tithymalus cyparissias. *Ib. f. 3.*" } ex Sternb.
" Pinus montana. *Ib. t. 14. f. 4.*"

? Palmacites incisus. *Schloth. Petrefactenkunde, p. 395. t. 15. f. 6.*

? Lepidodendron imbricatum. *Ad. Brongn. prodr. p. 86.*

From the roof of the Low Main coal seam, at Felling Colliery.

This species is no doubt identical with the plant figured by Count Sternberg, from the coal mines of Schatzlar and Swina, which he particularly characterizes by the rounded figure of the scars left by its leaves. This distinction, it must be observed, is only applicable to young branches of the species; in the old stems, with which Count

Sternberg seems to have been unacquainted, the scars are narrow lozenges, with a depression in their centre ; and are so like the figure of Palmacites incisus, given by Schlotheim, in his *Petrefactenkunde*, upon which M. Adolphe Brongniart founded his Lepidodendron imbricatum, that we can scarcely doubt their being the same. If we are right in this reference, the species has also been noticed in the slate-clay of Wettin, and Eschweiler.

It is readily known by its short compactly imbricated leaves, the form of which seems to have been ovate-acuminate, by the rounded scars on the young branches, and the narrow lozenge-shaped spaces, with a single central depression in the old ones.

In the specimen from which our figure was taken, two of the young branches were thickened, as if their leaves concealed axillary bodies. Should these be really the fructification of a Lepidodendron, we presume it will be no longer possible to admit the identity of Lepidostrobus, and that genus.

SPHENOPHYLLUM EROSUM.

————

Very rare in the shale above the Bensham coal seam, at Jarrow.

Whether this is either the S. truncatum, or S. dissectum, of M. Adolphe Brongniart, to which neither characters or references are assigned, we have no means of knowing; it is certainly distinct from all his other species. We beg, therefore, that our distinguished friend, whom we have no scruple in most conscientiously designating as the father of Fossil Botany, and as the only person that has hitherto viewed the subject, as Cuvier has the Fossil Animal Kingdom, with the eye of a man of science, and a skilful Naturalist,—we beg, we say, that he will not impute any deviation from his terminology, if into such we may fall, to disrespect; but rather, as we have already said, to mere unacquaintance with his materials. For differences in opinion, as to the inferences to be drawn from particular data, we feel that apology is unnecessary. No man is more capable than the learned Botanist, to whose name we are thus appealing, of appreciating the almost hopeless investigations of those who attempt to investigate the analogy of recent and fossil vegetable structure.

M. Brongniart refers Sphenophyllum to the family of Marsileaceæ;* but it seems to us, we confess, that this decision has been too hasty. It is true, that the leaves have the dichotomous veins of that family; and some analogy may, perhaps, be traced between their form and that of certain Marsileas. But when it is considered that the latter belong to a division of the vegetable kingdom, in which no such thing as verticillate leaves is known, and that all the Sphenophylla have their leaves most perfectly verticillate, it will at once be seen, that doubts may be reasonably entertained of the correctness of the approximation. An idea that Ceratophyllum has some relation to Sphenophyllum, has, probably, by this time, been abandoned.

While we thus differ from M. Brongniart, in regard to the families to which he has approximated Sphenophyllum, we are scarcely prepared to say to what else they are related. Perhaps, however, the following considerations may not be inappropriate.

* _Introduction to the Natural System of Botany._ _p._ 317.

There are no recent plants in which the veins of the leaves are dichotomous, except Ferns and their allies, and Coniferæ.

The veins of the leaves of Sphenophyllum are, in all cases, distinctly dichotomous, as is particularly seen in beautiful specimens of Sph. Schlotheimii, of which our indefatigable friend, Mr. Lonsdale, has shewn us specimens in the splendid collection of the Geological Society, to which they had been presented by the Rev. Mr. Skinner; therefore, Sphenophyllum is analogous, either to Ferns or Coniferæ, among recent plants.

While Ferns, and their allies, have constantly leaves with an alternate origin, Coniferæ have the leaves as frequently verticillate as alternate.

The leaves of Sphenophyllum are verticillate; therefore, Sphenophyllum is more nearly related to Coniferæ than to Ferns, and their allies.

This seems to us a legitimate conclusion, and it is strengthened by some circumstances that deserve notice. In the first place, the leaves of Sphenophyllum are dilated at the apex, like those of Salisburia, a genus of Coniferæ, and have exactly the same sort of veining: secondly, in the specimens above referred to, from Mr. Skinner, in the collection of the Geological Society, there seems to be a slight squamulose appearance at the base of each leaf, which, all Botanists will admit, would, if distinctly proved, be almost decisive of the question of the fossil belonging to Coniferæ; and thirdly, in the same beautiful specimens from Mr. Skinner, particularly in one numbered 16,916, from the Somerset coal field, the stem is distinctly marked with deep furrows, the ridges of which plainly correspond with the leaves. Now this is a character, so completely in accordance with that of the Yew, the Spruce Fir, and other Coniferous plants, that, taken together with what we have previously remarked, it leaves scarcely any doubt in our mind, that, Sphenophyllum was one of those plants, which in the ancient world represented the Pine tribe of modern Floras.

ASTEROPHYLLITES TUBERCULATA.

———

Ast. tuberculata. *Ad. Brongn. Prodr. No.* 6.

Bruckmannia tuberculata. *Sternberg Essai d'un exposé géognostico-botanique. fasc.* 4. *p.* xxix. *t.* 45. *f.* 2.

———

From the shale forming the roof of the Low Main coal seam, in Felling Colliery, near Newcastle.

Nothing more than fragments, such as are here represented, have been seen of this fossil; from which it is extremely difficult to form any clear idea of its nature. We have only portions of cylindrical stems, with internodia about twice as broad as they are long, and verticillate leaves, which are, however, so imperfectly preserved, that neither their outline, length, or number, can be judged of. They seem, however, to have been numerous. In some places, the central substance of the branch is laid bare, by the separation of what appears to have been a bark, of considerable thickness, in proportion to the whole diameter, and here the nodi are distinctly shewn to have been prominent. Such a specimen is represented at fig. 1. The only inference that can be safely drawn from this, seems to be, that the plant was *not* endogenous; if it had been, its cortical integument would not have separated in so distinct a manner, as it is evident that it did.

It might be suspected that it belonged to some species of Calamites, in consequence of its resemblance to the subjects of the two next plates, and more especially because specimens have been found of a size intermediate between the two; but there is no trace of the distinct parallel furrows, by which the stems of Calamites are to be recognized when their cortical integument is removed. Had the furrows been discoverable, it would certainly have been probable, that this fossil did belong to a Calamite, and C. approximatus might have even been named as the species; but the objection just mentioned, appears to render the supposition unsafe, until, what is not improbable, it shall have been discovered that very young branches of Calamite are destitute of the furrows.

This, undoubtedly, is very nearly the same as the fossil represented at fig. 2, of the first plate of Von Schlotheim's Beiträge zur Flora der Vorwelt, and compared, by that author, to the modern Hippuris vulgaris; but it would seem to have had shorter and more numerous leaves. That species,

with a few others, are placed by M. Adolphe
Brongniart at the end of his Arrangement of Fossil
Plants, under the genus Asterophyllites, with the
note, that they are perhaps the only traces of
Dicotyledonous plants in the coal measures, and a
suggestion that it is with modern Halorageæ,* or
Ceratophylleæ,† that they must be compared. The
latter part of this opinion being derived from an
inspection of specimens, apparently in fruit, such
as we have not seen, we are unable to judge of its
value : the former it is necessary to abandon entirely, in consequence of the discovery of those
undoubted Dicotyledonous plants already figured
in this work, under the name of Pinites ; to say
nothing of such others as it may be conjectured
belonged to the same great division of the vegetable kingdom.

These traces of axillary fructification are also
strongly dwelt upon by Count Sternberg, who refers the fossil to the heterogeneous assemblage
called, by some botanists, Naiades ; they are,
however, not represented in his figure.

* *Introduction to the Natural System of Botany, p. 57.*

† *Ibid, p. 176.*

CALAMITES NODOSUS.

C. nodosus. *Schloth.* *Petrefact, p.* 401. *t.* 20. *f.* 3. *Ad.*
Brongn. hist. des Végétaux Fossiles, 1. 133. *t.* 23. *fig.* 2—4.
C. tumidus. *Sternb. fasc.* 4. *p.* xxvi. according to Brong-
niart.
Volkmannia polystacha. *Sternb. fasc.* 4. *p.* xxx. *t.* 51. *f.* 1.

From the roof of the Low Main coal seam, in
Felling Colliery.

This belongs to a large and well known class of
fossils, of which the stems are more abundant in
the beds of the Carboniferous formation of the
north of England, than any others. They are often
found in close alliance with the coal itself, especially when thin layers of mineral charcoal are
discovered upon it.

In consequence of their abundance, and the
prominent feature they must have formed in the

E

most ancient Flora of the world, it is an object of great interest to determine whether they bear any near relation to existing tribes of plants, and what was their general appearance. The disjointed mutilated state of such specimens as have been found, throws peculiar difficulties in the way of such an inquiry, and have long rendered any positive conclusions, either as to their actual structure or botanical affinity, extremely unsatisfactory.

Much light has, however, been cast upon the investigation by the skilful labours of M. Adolphe Brongniart, by which we have profited in what we have now to remark, but to which we have also a few observations to add.

Calamites appear to have been branching plants, with hollow stems, and a distinctly separated wood and bark, often many feet in length, and readily disarticulating at their nodi. Their whole substance was extremely soft, so as to offer little or no resistance to pressure; and their internal cavity was apparently separated by phragmata, or horizontal partitions, at the nodi. The surface of their wood was marked with numerous parallel furrows, converging in pairs at the nodi, and there turning abruptly inwards, losing themselves in the phragmata.

That they were branching plants is evident, from the trace of scars at their nodi, from whence it is obvious that lateral ramifications have fallen, and also from such specimens as that figured in Mr. Artis's work, t. 2, in which a side branch is represented in connection with the main stem.

The hollow or fistular character of their stem, is demonstrated both by the power they possessed of yielding to pressure without material alteration of their outline, and by the occasional presence of fragments of ferns and other remains within them.

It is not quite clear whether wood and bark were both present, as has been above stated, or whether that coaly substance in which they are often found enveloped, does not in reality represent the whole thickness of the stem, the impressions of longitudinal furrows being casts of the inside of the stem, and not of the outside of a woody axis. M. Adolphe Brongniart has not overlooked this difficulty, without, however, being able to prove which supposition is the more probable. It may, however, be inferred, from his speaking of the *bark*, in describing these fossils, that he inclines to the opinion of their having had a distinct wood and bark. We concur in this opinion, for the following reasons. In the first place, the coaly matter that overlies the furrowed surface is apparently quite inadequate, from its thinness, to the support of plants, the diameter of which was frequently some inches, and the length generally many feet; secondly, the scars, whence branches have fallen, always indicate a thick solid mass within the coaly envelope; and, in the third place, if the coaly matter were the whole stem, there ought, *in all cases*, to be a fractured surface in impressions of the circumference of the nodi, because the phragmata which would, in that case, be continuous with the outer coaly invelope, must necessarily be broken through round all the circumference ; this *never happens*, as far as we have had an opportunity of observing.

We have said that the cavity of the stem was apparently separated by phragmata, or horizontal partitions, at the nodi ; and in speaking thus, we have adopted the expression of M. Adolphe Brongniart, and the common opinion upon the subject. At the same time, we feel considerable doubt of the accuracy of this view of their structure. It is not impossible, that what we call phragmata, may represent, in reality, the whole thickness of the wood, and that the open space that occupies the centre of these supposed phragmata, is all the cavity that existed in the stem. Supposing it should be demonstrated that there was both wood and bark in these plants, the latter opinion will be materially strengthened.

Some have supposed these fossils to have been analogous to reeds, whence the name of Calamites ; but there does not appear to be any solid ground for that opinion. M. Adolphe Brongniart has endeavoured to make out a close affinity with Equisetum, relying chiefly upon the resemblance in their furrowed stems, the lines of which alternate at their union at the nodi, and upon the presence of a sheath in his C. radiatus, analogous to that of Equisetum; and accounting for the more general absence of the sheath upon the well known, and, in Botany, incontestible principle, that the excessive developement of one organ (in this case the stem) often causes an abortion, or non-developement of a contiguous organ, (here the sheath.) But notwithstanding the ingenuity with which M. Brongniart has maintained his proposition, we confess there appear to us to be grave objections to it. He does not appear to have attached that value to the presence of wood and bark, which, in such an enquiry as this, is so important a circumstance in determining affinity. Nothing of the kind is known either in recent Equisetaceæ, or in any endogenous or monocotyledonous plants ; it is, on the contrary, strictly characteristic of exogenous or dicotyledonous plants. In Equisetum itself, nothing could produce such a clean separation of the inner and outer portions of the stem, as we find in Calamites; neither do we know of any recent endogenous plants in which it would happen ; it would, in all probability, not occur even in such as Aloe, which, although endogenous, have a distinct cellular integument.

We should rather consider Calamites as the remains of some dicotyledonous plants, the affinity of which, if any exist, has still to be traced.

One of M. Brongniart's arguments in favour of

a relationship with Equisetum, is derived, as has been stated above, from the discovery of a species of Calamite with the remains of a sheath, like that of the modern genus. It would be very desirable to ascertain whether that sheath is not of the same nature as the verticilli of leaves upon the specimens now represented, and which seem to be the leaves of a Calamite. We say seem, because, although we can scarcely doubt the fact, yet we know how unsafe it is, in this department of science, to make a single step without using the greatest caution. Although we have examined a fine series of specimens of this fossil, where the leaf-bearing branch is always associated with the stem, yet, as in no instance they have been found actually in conjunction, fig. 1. tab. 15., being the nearest approach to it that we have seen, we pause before we finally decide. Our specimens are too much mutilated for us to determine, with accuracy, either the form of the leaves, or their number, or the exact figure and manner of insertion of the young branches : the latter, however, always arise from a nodus, in the manner that is represented. In a part of one of our specimens a verticillus of leaves is depressed, and then resembles so very much the supposed sheath figured by M. Brongniart, that it is difficult not to suspect their identity.

Although we do not at present see that the discovery of these supposed leaves throws much light upon the affinity that Calamites bear to modern plants, yet it is obviously so extremely desirable to ascertain whether or not they really belong to the genus, that we trust our geological friends will neglect no opportunity of settling the question.

In the genus Calamites it is exceedingly difficult to determine what are called the species, even by the comparison of authentic specimens; and it is scarcely possible to doubt that a large number of them are merely different states of the same species. We presume this is the C. nodosus of Von Schlotheim, and A. Brongniart, although it does not retain the thick bark mentioned by the latter, as characteristic of that species. It is difficult to understand why C. ramosus is not also the same.

We think that Count Sternberg's Volkmannia polystachya, which seems to have been a little embellished by his artist, must be referable here.

17

ASTEROPHYLLITES GRANDIS.

———

From the roof of the Low Main coal seam, in Felling Colliery.

We find no record of this fossil, which is too imperfect to enable us to judge distinctly of its nature. It appears to have been a plant of considerable size, with numerous verticillate branches, and verticillate subulate leaves, arising from nodi, very remote from each other. Little more can be said about it, except that, in many respects, it may be compared with Calamites, from which it only differs in having its branches very imperfectly furrowed; a circumstance not unlikely to be due to the peculiar state of the specimen, and to its not being subject to such ready disarticulation, as is usual in that genus.

18

ASTEROPHYLLITES LONGIFOLIA.

A. longifolia. *Ad. Brongn. Prodr. No.* 4.
Brukmannia longifolia. *Sternb. essai d'un exposé géognostico-botanique. fasc.* 4. *t.* 58. *f.* 1.

From the shale in Jarrow coal mine.

It is probable, that this plant is of the same nature as Asterophyllites tuberculata, from which it differs specifically, in the much greater length of its leaves. At the same time, it must be remarked, that the specimens in our possession, although very perfect, do not exhibit any trace of the axillary bodies, said to exist in that species; and by which, indeed, the genus Asterophyllites is essentially characterized.

Count Sternberg refers it to Equisetaceæ, an approximation which it is difficult to reconcile with the existence of the axillary bodies. We suspect it is better, in the present state of our knowledge, to hazard no conjecture upon the subject.

Fig. 1.

BECHERA GRANDIS.

———

B. grandis. *Sternb. essai d'un exposé géognostico-botanique. fasc.* 4. *p.* 30. *t.* 49. *f.* 1.

Asterophyllites dubia. *Ad. Brongn. Prodr. No.* 10.

From the shale in the roof of the Low Main coal seam, in Jarrow Colliery.

M. Ad. Brongniart refers this to his genus Asterophyllites, among the doubtful species. We, however, think it better to preserve Count Sternberg's name, because it is scarcely to be doubted, that although it agrees with Asterophyllites in its verticillate leaves, it will prove, when better known, to be something widely different. This is indicated by its tumid joints, and deeply but widely furrowed stems, characters that are so distinctly marked, as to render it probable, that its texture was firmer, and its constitution different, from that of the other plants referred to Asterophyllites.

The leaves in this specimen are almost destroyed; but they appear, from Count Sternberg's figure, to be short, slender, pointed, and about four in a whorl.

No reasonable conjecture can be offered as to the affinity of this fossil and recent plants, until some more distinct information shall have been procured respecting its other states.

Fig. 2.

ASTEROPHYLLITES GRANDIS.

Ast. grandis. *Suprà, t.* 17.

From the roof of the Low Main coal seam, in Felling Colliery.

This represents the leaves of the fossil, figured at plate 17. They appear to have been about 14 in a whorl, very narrow, subulate, rather rigid, and perfectly distinct to the base. The stem was finely striated, and the joints not tumid.

LEPIDODENDRON OBOVATUM.

L. obovatum. *Sternberg. essai cah.* 1. *p.* 21. *t.* 6. *f.* 1. *and* 8.
f. 1. *a.* *Ad. Brongn. prodr. p.* 86.

From the roof of the Bensham coal seam at Jarrow Colliery.

This is evidently the same species as that found by Count Sternberg in the coal mines of Radnitz; and was, probably, a tree of considerable size. Specimens of a neighbouring species, L. aculeatum, were found in the same place, sixteen inches in diameter, at the lower end, (*Sternb.*); but these were even pigmies when compared with some that occasionally appear in the northern coal mines of this country. Portions of Lepidodendron have been there met with, in the roofs of the mines, from 20 to 45 feet long, and as much as four feet and a half in diameter.

It is perfectly distinguished by its obovate areolæ, of which the apex is rounded, the base tapering, the central ridge even and undivided, and the scar at the very apex of the areola bounded by a nearly circular outline.

CALAMITES ————; its phragma.

———

Specimens of this kind are very common in nodules of carbonate of iron lying among the shale in the coal measures; those now figured are from the roof of the Bensham coal seam, at Jarrow. We have several others from the coalfield of Barnsley, for which we are indebted to Mr. Edgar.

Their appearance is that of a circular flat body, with a crenated margin, from the re-entering angles of which run simple lines, converging towards the centre, but uniformly stopping short of it. In the specimens figured, they are but just within the margin; in others, they are equal to more than two-thirds of the whole diameter. In the latter case, they have very much the appearance of the recent fern, called Trichomanes reniforme, from which, however, they are distinguishable by their lines not being dichotomous.*

* A specimen of this kind formerly led me to suppose that I had met with a fossil instance, either of Trichomanes reniforme,

No one would suspect what these fossils are, from an examination of such specimens as those now figured ; they are proved by other instances to be nothing more than casts of the supposed partition, or phragma of the stem of some Calamite, of which two internodia have separated from each other. The crenatures are sections of the parallel striæ, and the converging lines are continuous with the furrows.

It is, perhaps, impossible, in the actual state of our knowledge of these plants, to tell whether the converging lines are horizontal vessels, or the ends of vertical plates, analogous to medullary rays. Supposing Calamites to have thin phragmata, the former would be most likely, although the absence of ramifications is a very unusual feature in veins ; but in the event of the supposed phragmata turning out to be disarticulated portions of wood, in that case, it may be expected, that they would indicate medullary rays ; to which the circumstance of several lines occasionally running

or of some species very nearly related to it. I took the crenatures for the remains of marginal fructification, and the lines for veins; an error from which I did not escape, until after a paper upon the subject had been read before the Geological Society. My mistake was pointed out to me by Mr. Robert Brown, who, when I first communicated to him my fossil, appeared to entertain the same opinion with myself; but who, after the paper was read, shewed me a proof of its being really the phragma of a Calamite. J. L.

side by side with each other, gives additional probability.

We have no means of offering even a conjecture as to the species to which these fragments belong.

21

CALAMITES ———; a crushed portion of the
stem (?)

———

From the roof of the Bensham coal seam, in
Jarrow Colliery.

This is no doubt a portion of a Calamites, which
has been struck perpendicularly so as to separate
it into many portions. Whether it was a young
stem, that had acquired no strength, or solidity,
or whether it was a part of its cortical integument
only, or whether, finally, it was an old stem,
which had, previous to the crushing, been so
much rotted, as to separate into several layers,
like the stems of many of our recent herbaceous
plants, it seems impossible to say.

We publish it chiefly, because any *fact* that
is connected with the illustration of the organization of this extensive fossil genus, is of too
much importance to be lost sight of.

The specimen from which the drawing was
taken, was about one-fourth larger.

22

CALAMITES MOUGEOTII.

Calamites Mougeotii. *Ad. Brongn. hist. des Végétaux fossiles, vol.* 1. *p,* 137. *t.* 25. *f.* 4—5. *Annales des sciences, vol.* 15. *p.* 438.

Copied from a drawing furnished for this work by Henry Witham, Esq. of a fine specimen in his collection, from the sandstone of the Edinburgh coalfield. The figure is one half the natural size.

In this instance we have the mode of branching peculiar to Calamites distinctly ascertained. The branches proceed from the nodi, gradually thicken as they lengthen, and afterwards taper off, so that the diameter of the two extremities is much less than that of the centre. In this respect Calamites resembles those recent Endogenous plants, which, like the Arrow-root, or some Cyperaceæ, emit subterranean stolones; and also differs from Equisetum, in which the young shoots are of nearly equal diameter throughout, even in the most gigantic species.

One of the branches on the left of our figure is divided, and seems as if the young lateral shoot emitted by it had a gradually attenuated termination. There is no trace of leaves upon any part of the specimen.

Although this is from the sandstone of the Edinburgh coal formation, yet it appears undistinguishable from C. Mougeotii, one of the few plants described from the new red sandstone of the Vosges by Mons. Adolphe Brongniart.

PEUCE WITHAMI.

This fossil was found in a sandstone quarry at Hill Top, near Ushaw, about four miles north west of the city of Durham. Unfortunately it was not *in situ*, but laid among the refuse of the quarry in a multitude of fragments, none of which were more than six inches in size; and which have been ascertained by Mr. Witham to have belonged to more than one species. The bed of sandstone is of the coal formation proper, and rather high in the series; a coal mine is worked beneath it, which is probably the Shield Row seam, as it is called in that division of the northern coalfield.

Another specimen, in the state of a rolled fragment, was found by Dr. Youens in a brook near Ushaw; and Mr. Witham picked another from a stone-heap by the road side.

The polished slices from which the drawings were taken, were communicated by the highly valued correspondent after whom we have named the species.

The transverse section (tab. 23.) offers to the naked eye a fibrous undulated surface, with several concentric lines of a deeper colour, at unequal and irregular distances (*a a a,* fig. 1.); but these, when examined by the microscope, are found not to be the concentric circles of an Exogenous plant, but to be merely waves, or slight alterations in the direction of the tissue of the fossil (see *a,* fig. 2.) Viewed by transmitted light beneath a magnifying power of 180, the appearance is such as is represented at fig. 2.; the tissue, which consists of the unequal elongated cellules of Coniferæ crossed by medullary rays, being displaced in many places by a deposit of inorganic semitransparent matter. The general character of this section is so much that of the Craigleith Fossil (Pinites Withami, tab. 2.) that it would be difficult to distinguish them. Like it, there is no trace of any concentric zones in a slice more than two inches across.

But in a longitudinal section (tab. 24.) the resemblance between these two entirely ceases. Instead of the finely reticulated structure of the walls of the cells of Pinites Withami, and which are peculiar to the genus Pinites, we have cells with a character entirely that of many Coniferæ of the present day; as, for example, of Pinus Strobus. The walls of the cells appear, under a power of 180 (tab. 24. fig. 1.) to have here and there upon their surface small roundish or oval areolæ lying either in single rows, or in two rows,

side by side, never occupying the whole of a cell,
but crowded irregularly towards one of its extremities, and often having themselves the appearance of having been pushed from their places by
violence. Still more highly magnified, as at
fig. 2., where a power of 500 is employed, many
of these areolæ are distinctly seen to have a minute central circle, which is sometimes opaque,
like the areolæ themselves, and occasionally transparent; when, if this happens upon an opaque
areola, it looks like a small hole. The greater
part of the areolæ are opaque, like the walls of
the cellules, but some of them are transparent;
and these latter may be observed, by the reflec-
tion of the light thrown upon them from the mirror
of the microscope, not to be plane, but to be
slightly convex. All this (with the exception of
the areolæ being often in *two* parallel rows upon
the walls of the cells, and *in contact* with each
other,) is so like that of Pinus Strobus, that no
reasonable doubt can be entertained of this fossil
being really a part of some tree analogous to recent Coniferæ. To distinguish it from Pinites,
which, as we have already shewn, can only be
considered an approximation to Coniferæ, the
name *Peuce* has suggested itself, that of Pinus
having been already applied to certain fossil cones
found in formations of a date much more recent
than the coal measures. *Peuce* will stand for the
generic title of all fossil wood that appears absolutely coniferous.

We are acquainted with no recent species in which either the areolæ of the tissue occupy two collateral rows upon the walls, or where there is no trace of concentric circles in so large a space as two inches across.

Fɪɢ. 1.

ASTEROPHYLLITES FOLIOSA.

From the roof of the Bensham coal seam, in Jarrow Colliery.

This was a tall branching plant, with long slender shoots, which were rather thicker at the base than at the apex. The nodi were scarcely at all tumid, and the internodia very slightly striated. The leaves grew 8 or 10 in a whorl, were perfectly distinct at their base, a little shorter than the internodia, and of a linear-lanceolate figure, with a slightly falcate direction. There seems to have been a midrib; but this is so imperfectly indicated, that nothing certain can be determined about it. No trace of fructification has been found.

At first sight, this seems to resemble some species of Asparagus; but, upon a more careful comparison, it will be found, that, while the branches of this are opposite, those of Asparagus are alternate; and that the leaves of the latter, although seemingly verticillate, are, in reality, fasciculate, and alternate ; or, in other words, grow in clusters from alternate points of the stem.

It is much more probable that this fossil, like the next, was of the same nature as our modern Stellatæ ;* from which we can only distinguish it in its actual state, by the want of sharp angles to the stem.

From Asterophyllites equisetiformis, it differs in its leaves not being more than one half the length ; from A. rigida, in the same circumstance, and in their being broader in proportion to their length ; and from A. diffusa, in their being much longer, and larger.

* *Introduction to the Natural System of Botany, p.* 202.

Fɪɢ. 2.

ASTEROPHYLLITES GALIOIDES.

———

From the shale of the Barnsley coalfield, communicated by Mr. Edgar.

Our specimen occurs among the remains of ferns, in fragments like that represented at fig. 2. No stem is visible. The leaves were in whorls of 10, had a lanceolate figure, were very acute at the apex, and had a distinct midrib, without a trace of lateral veins. Fig. 2 *a*, represents one of the most perfect of these leaves magnified; the specimen itself is of the natural size.

This is so very like some recent species of Galium, such as G. maritimum and murale, that it is scarcely possible to doubt its having been, at least, nearly related to them; and if any minute projecting points could be discovered upon the margin, or midrib, the identity would be almost established. In the mean while, we refer it to

the heterogeneous assemblage called Asterophyllites, because it is impracticable, from the imperfect state of our materials, to fix upon any generic
characters by which the verticillate leaved fossils
of this kind can be satisfactorily disunited.

LEPIDOSTROBUS ORNATUS.

Lepidostrobus ornatus. *Ad. Brongn. prodr. p.* 87.
Parkinson's Organic remains, vol. 1. *tab.* 9. *f.* 1.

From the shale of the Barnsley coalfield,
whence our specimen has been kindly sent, by
Mr. Edgar.

This, the original species upon which Mons.
Ad. Brongniart has chiefly founded his genus
Lepidostrobus, can scarcely be distinguished,
generically, from the fossils figured at our t. 10.
and 11., under the name of L. variabilis; and yet
it exhibits some striking marks of difference.   It
evidently was a cone, or strobilus, and consisted
of a number of woody plates, or scales, originating
in the surface of a central woody axis, of an elongated conical, or almost cylindrical figure, spreading nearly horizontally, turning backwards towards
the point of the cone at their extremities, and
enclosing organs of fructification.   From the scars
left upon the surface of this central axis, it is

certain that the scales, notwithstanding their apparent breadth, originated from a small roundish base; and that they had, also, a spiral arrangement. The recurved points of the scales seem to have formed rhomboids transverse with regard to the axis of the cone.

All these things are visible in the accompanying plate, in which fig. 1. is a portion of the upper end of a cone, of the natural size, lying imbedded in its stone, upon which the marks of the ends of the scales are impressed; and fig. 2. is the same portion of the cone separated from its bed, and magnified.

If the impressions of the origin of the scales upon the central axis be compared with those of any of the Pine tribe, in which the scales of the cone are deciduous, such as the Silver Fir, no one can fail to remark their general identity, as far as our means of comparison extend; but we can scarcely say that the resemblance goes further; and we certainly should not be justified in asserting, from what we at present know, that the structure of the organs of fructification enclosed between the scales of the cone, is the same as that of modern Coniferæ. In the conebearing genera of the latter, there are generally two short naked winged seeds, lying above each scale; and immediately upon the seeds, reposes the bracteal leaf that subtends each scale. But here it would seem, as if, between the scales, were enclosed several membranes, or leaves; and the seeds, of which we have fortunately discovered one, *in situ*, (see fig. 2. *a.*), were oblong bodies, nearly as long as the scales, and, most probably, altogether destitute of a wing. From the extremely brittle and mouldering state of the specimen we are now describing, we regret that we are unable to speak more exactly upon the subject. As these fossils are far from uncommon, we do trust that some of our friends will be able to discover the cones, not only in a still better state, but actually in connection with the leaves to which they belong. That the latter are well known, can hardly be doubted.

SPHENOPHYLLUM SCHLOTHEIMII.

Sphenophyllum Schlotheimii. *Ad. Brongn. prodr. p.* 68.
Palmacites verticillatus. *Schlotheim Flora der Vorvelt,*
t. 2. *f.* 24.

No. 16,916. *Mus. Soc. Geol. Lond.*

At page 41, we alluded to the existence, in the cabinet of the Geological Society, of fine specimens of this fossil, sent from the Somerset coalfield, by the Rev. Mr. Skinner. By the permission of the Society, we are enabled to publish the accompanying representation of these curious remains.

The stems appear to have been branched, and deeply channelled, the projecting ribs corresponding with the base of the leaves ; the internodia were rather shorter than the leaves. The leaves were whorled, and from six to nine in each verticillus ; they probably spread nearly horizontally : in figure they were exactly cuneate ; their apex was transversely truncate, and finely crenated, with a very slight appearance of an emargination in the centre ; the veins were dichotomously branched, and uniformly terminated in the sinuses of the crenatures of the apex ; the sides of the leaves were perfectly straight and undivided. At the base of the leaves, are here and there to be found obscure traces of what seem to have been scales ; but they are so imperfectly seen, that it is impossible to speak with confidence of their nature. No trace of any thing like fructification is discernable.

In the drawing, fig. 1. represents a portion of the fossil of its natural size, and fig. 2. a single leaf apart, and magnified so as to shew the veins distinctly.

We have already, in speaking of Sphenophyllum erosum, t. 13., adverted to the possibility of this fossil having more relation to Coniferæ than to any other recent family. In illustration of this suggestion, a drawing of a leaf of Salisburia adiantifolia, (fig. 3.) has been added to this plate, for the purpose of showing the great similarity in the arrangement of its veins. We confess we have no better arguments to offer upon this subject than those formerly adduced, but we still think them sufficiently powerful to render it improbable that Sphenophyllum belonged to Marsileaceæ, even if its approximation to Coniferæ should be rejected. Like all other questions in this department of science, nothing can positively be determined until fructification shall have been observed ; to the search after which we earnestly commend our readers.

Schlotheim asserts, that the leaves in his plant are always in sixes, and he so represents them : he also makes them have an entire roundish extremity. If these characters could be considered constant, his plant would be a different species from ours. We confess, however, that we distrust such supposed differences too far, to form a new species, when the general resemblance is so great. The conjecture of the learned German, that Sphenophyllum was of the Palm kind, seems by no means probable.

NŒGGERATHIA FLABELLATA.

Found, occasionally, in the shale, covering the Bensham seam of coal, in Jarrow Colliery. Tab 28. represents a leaf, one third the natural size ; tab. 29. is a detached leaflet of the size of nature.

Palms are so rare in the coal measures, that only one certain species, the Nœggerathia foliosa of Count Sternberg, of which a single specimen from Bohemia is in the Museum at Prague, has been discovered in Europe. We are so fortunate as to add another, which is referable to the same genus, but which is very distinct from Count Sternberg's plant.

A portion of a compound leaf, and a few scattered pinnæ, are all that have been met with. The leaf appears to have consisted of 6 or 7, or perhaps more, pairs of leaflets, which became generally smaller towards the extremity of the leaf; the midrib has not been distinguished. The most perfect pinnæ are cuneate, taper very much to the base, have a dilated, undulated, slightly lobed, crenated extremity, and appear to have

H

been flabelliform; others are narrower, and look like split portions of larger pinnæ, which, perhaps, they are.

That this is not a fern, is obvious from the veins not being distinctly dichotomous, but gradually separating, imperceptibly, as the pinnæ widen from the base, without any obviously marked point of divergence. Single pinnæ may, by this character, be safely distinguished from specimens of what Mons. Brongniart calls Cyclopteris digitata,* or similar plants.

The name Nœggerathia was given by Count Sternberg, in honour of Dr. Nœggerath, who has occupied himself specially in the study of fossil trees, and from whom much valuable information upon the subject is one day to be expected.

---

* Compare the figure of this in the Histoire des Végétaux fossiles, t. 61. bis fig. 2., with that of Salisburia adiantifolia, t. 27. fig. 3. of this work.

PINITES EGGENSIS.

WITHAM. *Observations upon Fossil Vegetables, p.* 37. *t.* 5. *figs.* 13 *and* 14.

For the preparation from which the annexed figure has been taken, we are indebted to Mr. Witham, by whom it was first described and figured in the work above referred to.

The bed to which this fossil belongs is not quite certain; but is supposed to belong " to the upper strata of the great Oolitic series." Mr. Witham obtained it " from the base of the magnificent mural escarpment of the Scuir of Egg," one of the Inner Hebrides.

In structure, it is obviously different from any of the Coal Coniferæ; its medullary rays appear to be more numerous, and frequently are not continued through from one zone of wood to another, but more generally terminate at the concentric circles; it abounds in Turpentine vessels, or lacunæ of various sizes, the sides of which are very distinctly defined; and here and there, rows of flattened tubes are found among the ordinary cylindrical woody tissue. These are distinctly visible in a cross section.

This forms one of the proofs, of which so many have now accumulated, that Dicotyledonous plants existed during the period of the Oolitic formation; and there seems to be no want of evidence to shew, that the earliest remains of land plants consist, more or less, of the highest orders of vegetables. It is, however, very remarkable, that, hitherto, no other kind of wood than the Coniferous should have been discovered in the older fossiliferous rocks, and that no positive trace of any other kind of Dicotyledonous tree should have been discovered earlier than the Lias.

We have not met with any recent plant of the same order, with the wood of which this can be considered identical.

Fig. 1. is a representation of a transverse slice the natural size; it is of a deep rich brown, which cannot be expressed without colour.

Fig. 2. is a small portion of the same, magnified 180 diameters; the larger oval, or round spaces, are the mouths of lacunæ; in the second zone of wood, from the bottom, are two rows of the flattened tubes, above alluded to; a third may be perceived, on the right of the third zone from the bottom.

31—36

STIGMARIA FICOIDES.

" Schistus variolis depressis; schistus variolis elevatis *Mo-rand, die kunst auf Steinkohlen zu bauen, t. 9. f. 3—4.*"

" Lithophyllum opuntiæ majoris facie. *Volkm. Siles. subterr. p.* 106. *t.* 11. *f.* 1."

" Cylindrus lapideus Byerleus compressior echinites lati-clavii maximi facie, acetabulis rotundis e puteis carbonariis prope Byerley in Yorkshire. *Petiv. gazoph. dec.* 2. *t.* 18. *f.* 11."

Phytolithus verrucosus. *Martin Petrificata Derbyensia, plate* 11, 12, 13. *Parkinson's Organic Remains, vol.* 1. *plate* 3. *f.* 1. *Steinhauer in Am. Phil. Trans. N. S. vol.* 1. *p.* 268. *t.* 4. *f.* 1—6.

Variolaria ficoides. *Sternb. essai. p.* 23. *t.* 12.

Ficoidites furcatus ⎫
——————— verrucosus ⎬ *Artis, Antediluvian Phytology,*
——————— major ⎭ *tab.* 3. 10. 18.

Stigmaria ficoides. *Ad. Brongn. in Mem. Mus. vol.* 8. *t.* 12. *f.* 7. *Prodr. p.* 88.

One of the most common, if not the most common, of the fossil vegetables of the Coal formation, is that now represented; which has, as its long list of synonyms indicates, been frequently before the subject of description. As the great multitude of its fragments, that are still every where to be found, assure us, that it must have formed a striking feature in primæval vegetation, we shall dwell at more than usual length upon its structure, and supposed affinities. But before we proceed to state our own notions upon the subject, we shall quote Mr. Steinhauer's ingenious paper in the first volume of the new series of the American Philosophical Transactions; which, although erroneous in some respects, is by far the best account of the plant that has yet appeared.

" The fossil which has received the name of Phytolithus Verrucosus from the ingenious author of the Petrificata Derbiensia, is by far the most common, and, perhaps, the most remarkable of this class. Woodward seems already to have collected numerous specimens, notwithstanding their bulk and comparative unsightliness; (Catalogue of English Fossils, vol. i. part ii. p. 104. vol. ii. p. 59, &c.) and Mr. Parkinson has exercised considerable, though fruitless ingenuity, in elucidating them. It might appear presumptuous, after the labours of men of such distinguished abilities, to obtrude to public notice any further remarks, had not these authors left abundant room for observation, which place of abode and inclination have enabled the writer to pursue, during a series of several years. Within this period we have collected several hundred specimens, worked many

from the bed of clay in which they were imbedded, and examined in quarries, on coalpit hills,
among heaps of stone by the road side, and in
various other situations, several thousand. The
Geological situation of this fossil is well known
to be the coal strata, in almost all which, as far as
the writer is enabled to judge, it is found. Its
geographical habitats in these strata, may be
partly collected from the works already quoted;
the specimens more immediately examined were
found in the neighbourhood of Fulneck, near
Leeds, or in the space included by the towns of
Leeds, Otley, Bradford, Halifax, Huddersfield, and
Wakefield; but we have also found it on the top of
Ingleborough, in the coal strata of Northumberland; abundantly in Derbyshire; at Dudley, in
Shropshire, and in the neighbourhood of Bristol.
With respect to mineralogical constituent matter,
it seems always to coincide with that of the stratum in which it is imbedded, with a slight modifi-
cation of density.

"It is most abundant in the fine grained siliceous
stone, provincially called *Calliard* and *Gannister*,
and in some of the coal *Binds* or *Crowstones*, which
have probably received this appellation from spots
of bitumen, or coal attached to these petrifac-
tions. It is rather less frequent in the beds of
scaly clay, or clay mixed with siliceous sand and
mica; very common but completely compressed
in the coal shales, or bituminous slate clay; of
occasional occurrence in the argillaceous iron

stone; not rare in the common grit, and upper thick beds of argillaceomicaceous sand stone or *rag*, and sometimes, though rarely, discoverable in the coal itself. Mr. White Watson, of Bakewell, had also in his collection which we examined, a specimen in the Derbyshire Toadstone or Trap, and we have also noticed it in the limestone behind the Bristol Hot Wells, at its junction with the sand stone. So immense, however, is the number of relics, that when the eye has been accustomed to catch their appearance, it is scarcely possible to walk a furlong in the districts where they are at home, without meeting them in one shape or another. The most perfect form in which this fossil occurs, is that of a cylinder, more or less compressed, and generally flatter on one side than the other, (Plate IV. fig. 1 and 2.) Not unfrequently, the flattened side turns in so as to form a groove. The surface is marked in quincuncial order with pustules, or rather depressed areolæ, with a rising in the middle, in the centre of which rising a minute speck is often observable. From different modes and degrees of compression, and, probably, from different states of the original vegetable, these areolæ assume very different appearances, sometimes running into indistinct rimae, like the bark of an aged willow, sometimes, as in the shale impressions, exhibiting little more than a neat sketch of the concentric circles. Mr. Martin suspected that these pustules were the marks of the attachment

of the peduncles of leaves; and his **Tab. XII.**
represents a specimen, in which he thought that
he had discovered the reliquia of the leaves
themselves. We have examined the specimen,
whence the drawing, which is extremely correct, was made; but are convinced, that Mr.
Martin was misled by an accidental compression,
in describing these leaves as being flat. Numerous specimens in gannister, in which the lateral
compression of the trunk is generally trifling,
place the assertion beyond a doubt, that the
fibrous processes, acini, spines, or whatever else
they may be called, are cylindrical; and small
fragments of these cylinders shew distinctly a
central line, (pith?), coinciding with the point
in the centre of the pustule. Convinced of the
existence of these fibres, we were soon able to
detect their remains forming considerable masses
of stone, particularly of Coal Bind on Wibsey
Slack, and at Lower Wyke, where their con-
torted figure imitates the figures of Serpulæ; but
it excited much surprise on examining the projecting ends of some trunks, which lay horizontally in a bed of clay, extending along the sou-
thern bank of the rivulet which separates the
townships of Pudsey and Tong, and which is
exposed, in several places, to find traces of these
fibres, proceeding from the central cylinder, in
rays through the stratum, in every direction, to
the distance of above twenty feet. Repeated observations, and the concurrent conviction of un-

prejudiced persons, made attentive to the phenomenon, compelled the belief, that they, originally, belonged to the trunks in question; and, consequently, that the vegetable grew in its present horizontal position, at a time that the stratum was in a state capable of supporting its vegetation, and shot out its fibres in every direction through the then yielding mud. For if it grew erect, even admitting the fibres to have. been as rigid as the firmest spines with which we are acquainted, it would be difficult to devise means gentle enough to bring it into a recumbent posture, without deranging their position. This supposition gains strength from the circumstance, that they are found lying in all directions across one another, and not directed towards any particular point of the compass.

" The flattened and sometimes grooved form of one side of the cylinder has already been noticed. Woodward already observed, that along this side, there generally, or at least, frequently, ran an included cylinder, which at one extremity of the specimen would approach the outside, so as almost to leave the trunk, while, at the other, it seemed nearly central. A reference to his Catalogue, vol. 1. part 2, p. 104, to Mr. Parkinson's Organic Remains, vol. 1. p. 427, and to Martin's Petrificata Derbiensia, l. c. will show how much this included cylinder has embarrassed those who have considered it with a view to the vegetable organ to which it owes its origin. In the spe-

cimens in Calliard, which have suffered little compression, but which are seldom above a few inches
in length, this body is generally nearly central;
perhaps, in no instance, perfectly lateral. In the
specimens in clay, from one of which we are able
to detach upwards of six feet, the flattened or
grooved side is invariably downward, and, consequently, the included cylinder in the position
which it would assume, if it had subsided at one
end, while the other was supported, or which
would be the result of its sinking through a
medium of nearly the same specific gravity with
itself, provided it was at one end rather denser
than at the other. It must be observed, that this
included body appears to have suffered various
degrees of compression, being sometimes cylindrical, which was evidently its original form, and
sometimes almost entirely flattened. In the coal
shale we were never able to detect a trace of its
existence.

" Besides these indications of organization, we
have met with several specimens, which, on being
longitudinally split, discovered marks of perforations or fibres, more or less parallel with the axis
of the cylinder, and, in some degree, resembling
the perforations of Terebellæ, in the fossil wood of
Highgate, and some other places. Whether these
configurations be owing to the organization of the
original vegetable, or to some process which it
underwent during its decay, seems impossible to
determine. The specimens examined, afforded no

opportunity of discovering a connection between these tubes, and either the internal cylinders or the external surface.

" Among the vast number of specimens examined, only one was detected, which appeared to terminate, closing from a thickness of three inches to an obtuse point. Two instances also came to our knowledge of branched specimens, in which the trunk divided into two nearly equal branches. So rare an occurrence of this circumstance would, however, rather induce the supposition that the original was properly simple, and that these were only exceptions or monstrosities. The size of different specimens vary greatly, but we have seen none under two inches in diameter ; the general size is three or four, and some occur, but with very indistinct traces of the pustules, even 12 inches across.

" From the above, it appears rational to suppose that the original was a cylindrical trunk or root, growing in a direction nearly horizontal, in the soft mud at the bottom of fresh water lakes or seas, without branches, but sending out fibres from all sides. That it was furnished in the centre with a pith of a structure, different from the surrounding wood or cellular substance, more dense and distinct at the older end of the plant, and more similar to the external substance, towards the termination which continued to shoot. And, perhaps, that besides this central pith, there were longitudinal fibres proceeding through

the plant, like those in the roots of Pteris aquilina. With respect to any stem arising from it,
if a root, or foliage belonging to it, if a creeping trunk, we have hardly ground for a supposition.

" If these points be assumed as ascertained, the
manner in which the reliquia were formed, is
easily accounted for. Annual decay, or an accu-
mulation of incumbent mud having deprived the
trunk of the vegetating principle, the clay would
be condensed by superior pressure around the
dead plant, so as to form a species of matrix. If
this took place so rapidly, that the mould had obtained a considerable degree of consistency before
the texture of the vegetable was destroyed by
putrefaction, the reliquium was cylindrical; if, on
the contrary, the new formed stratum continued
to subside, while the decomposition was going on,
it became flattened, and the inferior part might
even be raised up towards the yielding substance
in the inside, so as to produce the groove or crest,
as Woodward calls it, on the under side, in the
same manner as the floor in coal works is apt to
rise where the measures are soft, and the roof and
sides have been secured. While the principal
mass of the plant was reduced to a soft state, and
gradually carried away, or assimulated with
mineral infiltrated matter, the central pith, being
unsupported, would sink towards the underside,
and this the more sensibly where its texture was
most distinct, while its anterior extremity would,

probably, go into putrefaction with, and be lost in the more tender part of the plant. The mineral matter introduced would now form an envelope round the pith, where this resisted decomposition for a sufficient length of time, and when it was ultimately removed, if the surrounding mass was still sufficiently pervious, be also filled with argillaceous matter; or, if it was too much indurated, be left empty, which is the case occasionally. The epidermis or external integument of the vegetable, appears to have resisted decomposition the longest, as in many cases it has been preserved from putrefaction, in the manner necessary to change it into coal: its place more frequently, however, is occupied by a ferrugineous micaceous film. It, therefore, appears, that the original plants must have undergone a destruction by putrefaction, and the vacuities thus occasioned been very rapidly filled with mineral matter. This is evident from the reliquium, in its present state, exhibiting no minute traces of organization, nor any signs of bituminized vegetable matter, so frequent in siliceous and opaline wood, except in the epidermis, and from the close similarity which this substance bears with that of the surrounding stratum; whereas, in shells, &c. which have evidently undergone a very gradual lapidifying process, there is generally a very perceptible difference between the matter substituted and the surrounding mass.

" Several conclusions interesting to the science

of Geology, will readily be drawn. The formation
of these strata, from the deposit of water, is clearly
ascertained ; also, that the argillaceous strata in
question, must have been, when originally deposited, of nearly the same thickness as they now
are, as appears from the undisturbed position of
the vegetables of which they were once the bed,
and are now the tomb. On the other hand, the
shale of coal or slate clay, appears to have originated from a great number of successive depo-
sitions, which must have been of a very diluted
consistence, when vegetation became extinct in
the plants of which they now bear the impressions. All these strata must be supposed to have
been successively at no great depth from the surface of the water resting upon them, that these
plants might be supplied with air ; and the situation in which they are found, precludes the possibility of any motion of that sea sufficiently violent
to disturb the bottom. The general diffusion of
this, and several of the following species, strongly
suggests the belief, that all the coal strata through
which they are dispersed, owe their existence to a
similar origin."

Such were Mr. Steinhauer's opinions in 1818.

Count Sternberg, in describing it from the mines
of Radnitz, adverted to its affinity with Euphorbiaceæ, or Cacti.

M. Adolphe Brongniart, in his paper in the
Memoires du Muséum, objects to this affinity, and
suggested that it belonged rather to the family of

Aroideæ; an idea which, we confess, appears to us by no means well founded. In his last work, he refers the genus to Lycopodiaceæ, an opinion in which we are equally unable to concur. Mr. Artis adopts Count Sternberg's suggestion, that it was akin to Cacteæ.

Having prefaced thus much, we will next proceed to describe the accompanying plates; and then to see how far they corroborate or contradict Mr. Steinhauer's opinions.

Plate 31, fig. 1, represents the appearance of a nearly perfect specimen of this species, as it was laid bare by a fall of shale from the roof of the Bensham coal-seam in Jarrow Colliery. It is viewed from below, and, consequently, represents the under side. The central part, three feet in diameter, is concave; the whole surface being very distinctly covered with wrinkles, which, when attentively examined, are seen to be caused by depressed semicircular spots, compactly arranged in a spiral manner; in the centre of which is a roundish scar, to which a little fine coaly matter usually adheres. From this centre, arms, twelve in number, proceed on all sides; every one, when seen of length sufficient, dividing into two branches. The whole plant is flattened. As we recede from the centre, and approach the fore part of the arms, the circular tubercles, so well known, become more distinctly marked; and upon all the branches the leaf-like bodies remain attached. Upon several of the arms, the course of an internal central axis could be traced by a furrow, or depression in the fossil, as represented in the drawing.

Two other entire individuals have occurred, one of which having fallen whole from the roof, afforded an opportunity of examining the upper surface of the arms, which exhibited all the well known characters of the fossil; but the upper part of the centre itself was too much damaged to have its structure made out.

Plate 31, fig. 2, is an ideal vertical section, for the purpose of making more apparent what was the relative position of the parts when in situ.

Plate 32, is a diminished figure of a very fine specimen of a branch, showing that it was covered with tubercles, having an irregularly spiral arrangement. The bodies that proceed from these tubercles are too much crushed, to enable us to judge of their form.

Plate 33, is another specimen of the same kind as the last, with the tubercles more distinctly shewn; the spiral arrangement is here very much obscured.

Plate 34, is a portion of the arm from which the processes that arise from the tubercles have been cast; the spiral arrangement is here more distinct.

Plate 35, is a section of an imbedded stem from the Mountain Limestone district of Weardale, in the Sandstone of which formation it is abundant; it also occurs in the Limestone of the same group of rocks, near Wooler, in Northumberland, at

Birdy House, near Edinburgh, and in Fifeshire. The specimen shews that the axis was a woody core, communicating by means of woody elongations with the tubercles on the outside; this core has evidently contracted, since the plant was imbedded, and now lies almost loose in the cavity of the stem.

Plate 36, is a fragment of the stem in Ironstone, from Dysart in Scotland, from the Mountain Limestone formation; the specimen had been irregularly pressed and bruised before it hardened; and its core is seen to be very excentrical.

From all that has now been adduced, it would seem that the following inferences may be drawn.

1. *That Stigmaria was a prostrate land plant, the branches of which radiated regularly from a common centre, and, finally, became forked.* What the nature of the centre itself was, it is difficult now to conjecture; we only know, that it really belonged to the system of the stem, by the scars still remaining upon its surface. Perhaps, what seems in the fossil state to have been a continuous homogeneous cup, or rather dome, may, in reality, be nothing more than the arms squeezed into a single mass where they came in contact, their lines of separation being no longer traceable. If a domed centre was the natural character of the genus, it was unlike any thing we now have; but is it not possible, that the domed appearance may have arisen from the plant when imbedded having been growing from the summit of a small

rounded hillock? Of its roots, nothing is known; but if small, and proceeding, as they no doubt did, from the very centre of the dome, they would, necessarily, be broken away with the mass of shale which separated from the plant, when it was left hanging in the roof of the coal mine.

2. *That it was a succulent plant.* Of this the compression of the stems seems to offer a proof; to which may be added, the frequent excentricity of its *core*, or woody axis, which may have been owing to some inequality of the pressure to which it was subjected. But if this evidence is thought insufficient, at least, the specimen, represented at Tab. 35, which is by no means an uncommon state of the fossil, must be considered a strong corroboration of the opinion. It is well known, that if recent succulent plants, that are old enough to have formed a woody axis, are placed in a situation in which decay takes place, the soft parenchyma of the bark, and of the interstitial medullary rays of the wood deliquesces, and leaves the woody axis loose within the still undecayed external portion ot the bark. We have had occasion ourselves to notice this in Cactus Pereskia, and old stems of Opuntia. If this axis be examined, it will be found covered with woody prolongations, which were the channels of communication between the buds and the wood. A structure more analogous to that of Stigmaria, can scarcely be wished for.

3. *That it was a Dicotyledonous plant.* This

K

may be inferred from the existence of a central woody axis, from which the bark has separated. If it were a Monocotyledonous plant, no such separation would have taken place, and no Cryptogamic plant has a solid central axis, with a distinct cortical integument.

4. *That the tubercles upon the stem are the places from which leaves have fallen.* This is proved by the great regularity with which they are arranged upon the older stems. Roots never proceed from a stem with any kind of symmetry: hence, Steinhauer's conjecture, in this respect, is inadmissible.

5. *That the leaves were succulent and cylindrical.* There is, no direct evidence of this; but it seems probable that such was the case, from the crushed and shapeless state of the flat specimens, and from Steinhauer's observations on such as were embedded without compression. Mr. Artis's figure represents them as somewhat cylindrical; but we have never been able to discover an instance of the forking that he speaks of, and figures in some specimens of leaves. With regard to their length having been as much as twenty feet, as Mr. Steinhauer states, we think there must have been some mistake in the observations upon which that report was made.

What the analogy is, that this curious plant bore to species of the present day, it is, perhaps, impossible to demonstrate. That it did not belong to Aroideæ, as Brongniart once surmised, is apparent. Was it a Lycopodiaceous plant, allied to Isoetes, as that ingenious author now conjectures? We think assuredly not. Indeed, we are so much at a loss to discover in what the resemblance consists, that we think that opinion must be abandoned, especially as it can scarcely now be doubted, that Stigmaria was Dicotyledonous. The only point of structure that seems to us to render it probable that it was a Lycopodiaceous plant, is the bifurcation of the branches, a character which, unless accompanied by other evidence, cannot be considered of great importance.

We must look, then, among succulent Dicotyledons for an analogy; and Euphorbiaceæ, Cacteæ, and Asclepiadeæ, at once suggest themselves. In fact, if we compare the axis of Stigmaria with that of Cactus Pereskia, the resemblance is most striking; but then it is probable that the axis of any succulent Dicotyledonous plant would exhibit the same appearances, so that the loose axis of Stigmaria would indicate a relation to Euphorbiaceæ, or Asclepiadeæ, as well as to Cacteæ.

The Stapelias of the Cape of Good Hope, or the Carallumas of India, have a trailing habit, similar to that of Stigmaria; but, it must be confessed, this is but a rude kind of analogy. We should rather incline to the belief, that it is between Euphorbiaceæ, or Cacteæ, that the Botanist has to decide, if an existing analogy must be found; and if we take the former in preference, it is rather because their fructification is so

minute as to be easily lost or overlooked in a
fossil, and that they have a greater tendency to
the development of leaves; while Cacteæ, on
the other hand, have so highly developed a flower,
that it could not be overlooked, or lost; besides,
the fructification of Euphorbiaceæ is deciduous,
that of Cacteæ persistent. We have left the
succulent families of Crassulaceæ and Ficoideæ
out of the question, because no existing genera
of those orders approach Stigmaria in the smallest
degree.

We presume, that the specimens of Stigmaria,
here represented, are all referable to one and the
same species, in different states; at least, we can
discern no characters that we dare trust to distinguish them. Of Mr. Artis's species, Ficoidites
furcatus is from near the extremity of a branch,
with the leaves on; F. verrucosus is a branch
that has lost its leaves; and F. major seems to be
the lower part of the same, where the tubercles
are more deeply impressed.

PECOPTERIS ADIANTOIDES.

From the Bensham Coalseam in Jarrow Colliery.

With this beautiful species we commence our illustrations of Fossil Ferns, by far the most remarkable of the tribes that formerly covered the crust of Great Britain, and the most susceptible of positive determination. If the species that are found in a fossil state are not capable of being reduced to the genera of modern Botanists, this is of little importance, when we consider how artificial those genera are, how bootless is the labour of attempting to reduce the fossil species to those of the existing æra, and how probable it is that those principles of determining genera by the arrangement of the veins, and by the divisions of the leaves, which Adolphe Brongniart has so judiciously pointed out with regard to fossil species, may be sooner or later adopted by Botanists in the recasting the genera of modern ferns.

The species now before us belongs to a genus called Pecopteris, which is characterized by the

L

leaves being once, twice, or thrice pinnate, and by the leaflets having a perfect midrib, from which forked veins proceed more or less at right angles with it.

With reference to modern ferns, it may be compared, as our valued friend Dr. Hooker reminds us, with Adiantum obtusum, from which its venation distinguishes it, and also with certain Aspidia, Polypodia, and Asplenia. Compared with fossil species, it is so like Pecopteris oreopteridis, a plant found in the slaty clay of Manebach and Radnitz, that we can find nothing, except its *being twice the size* of that species to distinguish it.

It appears to have been bipinnate, with its leaflets nearly of equal size, adherent to the rachis by their base, of ten or eleven pair with an odd one, each being oblong and entire, with a very rounded apex.

NOTE.

We are requested by Mr. Witham to say, that the Pinites Eggensis, figured at t. 30 of this work, was communicated by Mr. Nicol, its original discoverer.

38

PECOPTERIS HETEROPHYLLA.

From the high main Coalseam in Felling Colliery.

It was found in great abundance in one small district of that seam, but has not been met with any where else.

This species is so nearly the same as Pecopteris aquilina, figured by Schlotheim from the Coal measures of Manebach and Mandflech, that it may almost be considered the same. It appears, however, to differ essentially in the leaflets being narrower, more tapering to the point, and much longer; and also, as far as we can judge from our specimens, in that species having been of a more gigantic habit.

If compared with recent species, we would at first sight pronounce it to be a Pteris, and even Pt. caudata, a plant that occupies at the present day the same station in North America, that Pt. aquilina holds in Europe; and upon comparing the fossil with the recent plant, this idea is so much strengthened, that we cannot doubt that their nature was the same. Nevertheless in this, as in all similar cases, close resemblance proves, upon very accurate comparison, not to be the same as identity; for in the fossil the lateral veins are all simple, in the recent Pterides that resemble it the veins are all dichotomous.

The fossil seems to have been a pinnate plant, with its lower pinnæ deeply pinnatifid into linear almost falcate segments, traversed by a single midrib, from which arise numerous simple veins; the upper pinnæ entire, and nearly as long as those with the pinnatifid structure, from which they abruptly change without any pinnatifid appearances upon themselves. Each pinnatifid pinna is about $2\frac{1}{2}$ to 2 inches long, and its segments about one-sixth of that length; or if the pinnæ are longer, the leaflets are in the same proportion.

Fig. 1. is the natural size; fig. 2. is magnified a little.

39

SPHENOPTERIS CRENATA.

From the Bensham Coalseam in Jarrow Colliery.

To this genus Sphenopteris are referred all Fossil Ferns, with twice or thrice pinnated leaves, the ultimate leaflets of which do not adhere to the rachis by their whole base, and are traversed by one or two principal veins in each lobe.

The subject of this plate is closely allied to Sphenopteris tridactylites, from which it differs in the lobes of its leaflets being shallower, and never toothed, or in any degree divided. S. hymenophylloides is another fossil species with the aspect of this; but it has the partial rachis bordered with a membranous continuation of the base of the leaflets, so as scarcely to come within the character of Sphenopteris, and is, moreover, an Oolitic species.

This has in some respects the aspect of modern Dicksonias, but we know no species with which it can be strictly compared.

The principal rachis seems to have been disproportionably thick, for the size of the leaflets, and of the partial rachis, which is in no way bordered with membrane, but distinctly rounded. The leaflets were oblong, rather dilated at their base, and divided on each side into about six entire crenatures, which become gradually smaller towards the apex. To each leaflet there is one principal midrib, from which one single vein diverges into each crenature, losing itself before it reaches the margin.

40

ODONTOPTERIS OBTUSA.

Odontopteris obtusa. *Ad. Brongn. prodr. p.* 60. *Histoire des Végétaux fossiles, p.* 255. *t.* 78. *fig.* 3. 4.

The specimens from which the accompanying figure was taken, were communicated by Professor Buckland, from a Coal Pit belonging to Archdeacon Corbett, at Leebotwood, four miles from Church Stretton, and nine miles from Shrewsbury.

It is evidently the same as the plant found by Mr. Brard in the Coal measures of Terrasson. The specimens we have seen are, like those examined by Adolphe Brongniart, mere fragments; they nevertheless appear distinct from any of those with which we are acquainted in a more perfect state.

What is represented may have been the extremity either of a pinnatifid leaflet, or of a pinnated leaf; there is nothing in the specimen to show which. The lobes are oblong, rounded at the end, nearly three times as long as broad, reckoning from the common midrib to their apex, and gradually diminishing in size till they terminate in one broad blunt lobe, at least twice as large as any other. The veins of each lobe are once or twice dichotomous, but obscurely marked, and all originate in an indistinct common midrib, passing through the axis of the leaflet; there is no midrib to the lobes. By this arrangement of the veins, Odontopteris is characterized as a fossil genus.

41

NEUROPTERIS CORDATA.

Neuropteris cordata. *Ad. Brongn. Hist. des Végétaux fossiles,*
p. 229. *t.* 64. *f.* 5.

Like the last, from Leebotwood Coal Pit, and
communicated by Professor Buckland. It has also
occurred in the mines of Alais and St. Etienne in
France.

It appears to have been a fern of large size,
judging from the unusual dimensions of the leaflets, from a fragment of a rachis represented at *b*,
from crushed remains of other portions of a still
wider rachis, and from a flattened cast, three
inches wide, of a fossil found in its vicinity, which
has been the stem of some arborescent fern. The
latter is unfortunately in too imperfect a state to
be represented or even described.

The leaflets have generally no attachment to the
rachis, but are found lying loose in the shale;
from which it may be inferred that they were not
suddenly buried in consequence of some convul-

sion, but were probably shed by the tree at the period when they naturally disarticulated. They are from 3 to $4\frac{1}{2}$ inches long, of an oblong figure, acute at the apex, cordate at the base, very black and shining, and with no other midrib than what is produced by the united bases of their veins, which diverge from the axis of the leaflets, forming curved dichotomous lines that reach the margin. The margin itself is perfectly entire. At irregular intervals the veins are more than usually well marked; we know not whether this is accidental, or characteristic of the species.

Among the large leaflets are found others of a nearly circular form, not more than half an inch in diameter, and having veins radiating and dichotomizing with great regularity from their base, which is cordate. These, with the existence of which, as well as of the evidence of the gigantic habit of our fern, Brongniart was unacquainted, were doubtless the diminutive basal leaflets of one of the pinnated divisions of the leaf, such as are found upon the recent Pteris atropurpurea, and upon the fossil Neuropteris heterophylla. They are represented at *a a a* in the plate.

The leaflets of this plant have very much the aspect of the wild Osmunda regalis, which has also a tendency to the production of small leaflets at the base of the larger. But with this *primâ facie* similarity, all comparison ceases, for the recent plant would be a Pecopteris if found in a fossil state.

CAULOPTERIS PRIMÆVA

———

By permission of the Geological Society, we are enabled to publish the annexed representation of the only well defined specimen that has yet been found in the Coal measures, of what was certainly the stem of a tree fern. It was discovered in the Coal mines of Radstock, near Bath, and was originally pointed out to us by Mr. Lonsdale.

If it be compared with the recent stems of such a Fern as Dicksonia arborea, or any of the West Indian Cyatheas, in which the scars of the leaves are very much elongated, it is impossible not to perceive their striking resemblance, particularly when some of the fibrous matter that clothes the recent stems has rotted away.

The specimen before us is a compressed fragment, with both its sides nearly equally perfect. Its surface is depressed into shallow sinuous furrows, that form very elongated rhomboidal ridges, the upper part of which is marked with a long oval scar, very much broken at the edges, and on the surface; these scars are from three to four

times as long as broad, and are disposed in a spiral
manner, about four scars of each spire occupying
one of the compressed sides of the fossil; so that
it may be supposed that eight leaves went to the
making up of one complete turn of the spire when
the plant was growing. Over all the intervals
between the spires, in patches of various sizes,
extends a sort of coaly covering, looking like a cortical integument, and having a great number of
very minute transverse cracks. There is no trace
of any internal organization.

We have already, in the preface to this volume,
pointed out the error of considering the fossils
called Sigillarias, as the remains of Tree Fern
stems; the subject of this plate will make this
sufficiently apparent, if compared with any of the
known species of Sigillaria. What the latter may
have been, it is, perhaps, impossible to determine;
we shall, however, in the next part of this work,
endeavour to show what their structure was, and
to point out such analogies as can be detected between them and recent plants.

Fɪɢ. 1—2.

CYPERITES BICARINATA.

———

From the Leebotwood Coal Pit, communicated by Professor Buckland.

It is a very remarkable circumstance that we have no published evidence of the existence of Glumaceous Monocotyledons* in the Coal measures, for the fossils called Poacites appear to have been narrow Monocotyledonous leaves, not belonging to the tribe of Glumaceæ.

In figuring this under the name of Cyperites, we do so rather from inability to match it with the leaves of any other family, than from any conviction that it really belongs to Cyperaceæ.

In all Gramineæ, Palms, or narrow-leaved Monocotyledons, that we have noticed, there is uniformly a midrib, with which the other veins are more or less parallel ; and it is for the purpose of comprehending all such fossil leaves that the genus Poacites has been constructed. It is only in the

---

* *Introduction to the Natural System of Botany, p.* 292.

order Cyperaceæ, and in the genus Cyperus, that
we have remarked any deviation from this kind of
structure. In some species of Cyperus, the midrib becomes so indistinct and flat, particularly on
the upper surface of the leaf, that in a cast
it would be altogether obsolete; while at the
same time two lateral veins become unusually
highly developed. This happens in Cyperus pro-
cerus, *Roxb.* (Wallich's Cat. No. 3329,) and is
the characteristic structure of the present fossil.

It is found in short fragments, lying with Neuropteris cordata, appears to have been a long,
narrow, ensiform leaf, and of a rigid texture. It
has no visible midrib; but has two parallel, simple
ribs, depressed on the upper side, and prominent
on the lower, rather nearer to the margins than to
each other, and each accompanied by two very
fine veins, of which the two inner are rather more
remote from the main ribs than the outer. No
trace of any other veins can be detected in the specimens that have come under our inspection.

The want of midrib, and the presence of lateral
veins, will therefore be the distinguishing characters of the genus Cyperites.

Fig. 1. represents the fossil of its natural size.

Fig. 2. is a magnified representation, to show
the secondary veins.

Fɪɢ. 3.

LEPIDOPHYLLUM INTERMEDIUM.

From the Leebotwood Coal mine.

We find nothing among the fragments or the specimens communicated by Professor Buckland, to which this leaf can be supposed to have belonged. It is of the same nature as that represented at t. 7 of this work, and is as it were intermediate between that species and Lepidophyllum majus.

The strong midrib, peculiar outline, want of lateral veins, and apparent texture of these fossils, seem to point out a greater degree of affinity with the Coniferous genus Podocarpus, than with any thing else among recent plants. At least, no difference can be discovered upon comparison of the two, except such as may indicate a want of specific identity.

CYCLOPTERIS BEANII.

For an excellent drawing and description of
this very remarkable plant, we are indebted to
Mr. William Williamson, of Scarborough, for a
careful account of its usual appearances to Mr.
Dunn, and for the examination of a specimen to
Mr. Bean, of the same place. From these materials we are enabled to draw up the following
account.

It was found by Mr. Williamson, Sen. in the
deposit of plants in the Upper Sandstone* and
Shale (of Phillips,) at Gristhorp Bay, where only
two specimens have, however, been detected.
We ought, therefore, perhaps, to have called it
Williamsoni, rather than Beanii; but we had
given the name to the specimen, obligingly sent

---

* " My father has some doubt, whether it is the upper or lower
Sandstone and Shale in which the plants are found at Gristhorp Bay. From circumstances connected with the neighbouring strata, he seems to consider it an uplifting of the
latter." *W. Williamson, Jun.*

us by Mr. Bean, before we knew by whom it was discovered.

The plant appears to have grown to a considerable length; Mr. Williamson's drawing measures 18 inches; it consisted of a flexuose axis, gradually tapering from the base to the apex, and having four or five longitudinal furrows distinctly impressed upon its lower part. From this axis, and almost at right angles with it, spring numerous imbricated leaflike bodies, each of which is of an oblong figure, broader on one side than the other, decreasing from an inch and a quarter to less than half an inch in length, perfectly entire at the margin, and marked with five equal-sized veins that radiate from the base to the circumference in a flabelliform manner, dichotomizing so as to fill the margin as full of their ramifications as the base. The dilatation of one side of these bodies, which is the cause of their obliquity, is in all cases towards the extremity. Their stalk is not a mere lateral expansion of the border of the axis, but originates from across the axis, as the leaflets of Palms from their petiole.

Such being the structure of this plant, it becomes an enquiry of some difficulty to determine how to name the parts that have been described, and thus to judge of the real nature of the fossil. It looks at first sight like a pinnated leaf, of which the axis is the petiole, the lateral foliaceous bodies the leaflets. But we only know one tribe of plants in which the leaflets are set on across the petiole, and that is the Palm tribe; and we have no modern instance of a flexuose petiole in Palms, nor of a form of leaf like this decreasing gradually from the base to the apex. Is it the pinnated leaf of a Cycadeoidea? among the remains of other species of which it was found; certainly not: for the setting on of the leafy parts is at variance with that of Cycadeæ, and the veins are dichotomous. Is it a pinnated Fern? Its veins agree, but the setting on of the leafy bodies again renders this improbable.

We believe it to be not a pinnated, but a creeping Fern, such as many Hymenophylla and Polypodiums found in tropical countries at this day. The flexuose axis appears to have been a creeping stem, or rhizoma, and the foliaceous bodies to have been leaves growing from that creeping stem, small, when young, at the upper end, and full grown only towards the base of the specimen. If this view of the subject is correct, then this will have been a Cyclopteris, of which one other supposed species, the curious C. digitata, has already been found in the Oolite.

45

SPHENOPTERIS AFFINIS.

Communicated by Mr. Witham. It occurs in a fine hard blue mountain limestone in the great Lime quarries near Gilmerton, a little south of Edinburgh, where it is associated with finely preserved remains of other ferns, Lepidodendra, Lepidostrobi, Stigmariæ, &c.

This beautiful species is nearly related to the subject of the next plate, from which it differs principally in being smaller in all its parts, with shorter lobes to its leaflets, and a larger number of divisions. It may be a mere variety of it; but if so, it is too well marked to be omitted.

The leaf was bipinnate, the leaflets being deeply pinnatifid into five segments, each of which is divided into from three to five linear obtuse segments, which are broadest at the upper end, and marked with from one to three parallel veins.

46

SPHENOPTERIS CRITHMIFOLIA.

From the roof of the Bensham Coal main in Jarrow Colliery.

This is so like Sph. Artemisiæfolia, a species described and figured by Count Sternberg, and by Ad. Brongniart from the Newcastle Coal field, that we were at one time disposed to believe them identical. But, upon comparing this with the magnified representations of the latter Botanist, we have come to a different conclusion.

It was a bipinnate fern, with pinnatifid segments, the four or five divisions of which are split into about three long linear obtuse lobes, each of which has from one to three veins. In Sp. Artemisiæfolia are the contrary; the segments have from five to nine divisions, the lobes of which are broader, far less deeply split, and marked with from five to seven veins. In this they agree, that the main petiole is forked near its middle.

47

SPHENOPTERIS DILATATA.

From the roof of the Bensham Coal seam, in Jarrow Colliery.

At first sight, we took this for the Sphenopteris obtusiloba, of Brongniart; but an attentive examination of its veins shews that it is not only not that species, but that it would belong to the genus Odontopteris, to a species of which, O. Schlotheimii, it nearly approaches, if its leaflets were not contracted into a sort of petiole at their base.

It is distinguished by the final divisions of the segments of the leaves, or the leaflets, being either entire, or two-lobed, or even three-lobed; the lobes being, in all cases, sensibly dilated at the apex, and the divisions themselves placed widely apart, and contracted into a sort of petiole at their base.

48

SPHENOPTERIS CAUDATA.

From the roof of the Bensham Coal seam, in Jarrow Colliery.

Apparently this is very nearly the same as Sph. Virletii, than which it is far smaller. It occurs in fragments so much broken that it is impossible to say what has been the degree of division of the perfect leaves. All that we have seen are pinnated portions, with long, narrow, taper-pointed leaflets, pinnatifid at the base, crenated at the apex, both segments and crenatures being rounded and one-veined.

We should conjecture, from their general appearance, that these were fragments of some very decompound leaf, of which they are merely the terminal portions.

49

NEUROPTERIS LOSHII.

Lithosmunda minor, &c. *Lluid. lithophyl. Brit. ichnogr. p. 12.*
t. 4. f. 189.

Neuropteris Loshii. *A. Brongn. prodr. p. 53. Hist. des veg.*
foss. p. 242. t. 72, 1, and 73.

In shale, from Felling Colliery. This specimen is an impression of the under side of the leaf.

This fern first appears in Lloyd's rare Lithophylacium, published in 1760, in which work a lower pinna, without the large terminal pinnule, is figured from the Coal mines, near Gloucester. It has since been found in those of Yorkshire, and Northumberland, in the North of France, and even in Pennsylvania.

It seems to have been a fern of considerable size, the part now represented being only a portion of the upper extremity of a bipinnate leaf. Towards the base, the leaflets of the pinnæ are all nearly equal in size, of an oblong or ovate figure, diminishing to the apex ; but, towards the upper extremity, the terminal leaflet is much larger than the rest, and of a more elongated figure. The difference in form between the lower and terminal leaflets of this species, will give an idea of the relation that the two forms of leaflets in N. cordifolia bear to each other.

50

NEUROPTERIS SORETII.

N. Soretii. *Ad. Brongn. Prodr. p.* 53. *Hist. des Végétaux foss.* 1. 244. *t.* 70. *f.* 2.

This very distinct Neuropteris, from Felling Colliery, does not appear distinguishable from a species found in the Anthracite of Savoy, by M. Soret, after whom it has been named.

It occurs in broken fragments, appearing to have belonged to a bipinnate leaf, of which some of the pinnules bear as many as thirteen pairs of leaflets. Of these the terminal one is, in the specimen now represented, not much larger than the lateral ones ; but in Brongniart's figure it is very considerably larger, and differently formed; this, however, is as likely to be owing to different portions of the leaf of the same species having been

N

preserved, as it is to indicate a specific difference. The lateral leaflets are very exactly oblong, obtuse at each end, and do not overlap each other, except towards the extremities of the pinnules.

51

NEUROPTERIS ACUMINATA.

Filicites acuminatus. *Schlotheim Petrefaktenkunde, p.* 412. *t.* 16. *f.* 4.

Neuropteris smilacifolia. *Sternb. tent. fl. prim. p.* xvi.

europteris acuminata. *Ad. Brongn. prodr. p.* 53. *Hist. des Végétaux foss.* 1. 229. *t.* 63. *f.* 4, *copied from Schlotheim.*

From Felling Colliery.

Except in its leaflets being less cordate, we do not distinguish this from the very rare fossils found in the Coal measures of Kleinschmalkalden, and figured by Baron von Schlotheim. Those specimens and ours are both in nearly the same state, so that it is impossible to say to what kind of stem or rachis they belonged.

M. Brongniart points out their general resemblance to some species of Lygodium. If they really belonged to such plants, and the very remarkable similarity between them and the barren leaflets of L. microphyllum makes such a conjecture not improbable, the species must have been a climber, and the part now represented a lateral pinnule of a much branched axis. In recent Lygodia the base of the leaflets, when cordate, varies so much from that to a merely ovate form, especially in L. microphyllum, pubescens, and the like, that we do not doubt that circumstance to be unimportant as a specific distinction.

The principal objection to this having been a Lygodium, appears to us to consist in the great breadth of the petiole of the Fossil, and its slender character in the recent species. In the specimen now before us, (the only one that has been discovered,) the petiole is, at its widest part, about two lines broad, and looks as if it had been flat. But the venation of the leaflets, wide near the middle, and gradually becoming more and more dense towards the margin, in consequence of the dichotomizing of the veins, is altogether that of Lygodia.

52

NEUROPTERIS GIGANTEA.

Filicites linguarius. *Schloth. Petrefaktenkunde, p.* 411. *Ejusd. Flor. der vorw. t.* 2. *f.* 23.

Osmunda gigantea. *Sternb. Flor. der vorw.* 3. *p.* 29. 33.

Neuropteris gigantea. *Sternb. Tent. fl. primord. p.* 16. *t.* 22. *Ad. Brong. prodr.* 54. *Hist. des Végét. foss. t.* 69.

From the Coal measures of Saarbruck, Eschweiler, Wettin, and Kleinschmalkalden, according to Schlotheim, and of Schatzlar, according to Sternberg, and also of Newcastle, our specimen having been procured from Jarrow Colliery.

With the modern genus Osmunda, to which it has been referred, it does not appear to have any thing in common except the form of its leaflets. Its stature was probably considerable, the fragment figured by Sternberg having a rachis so

N 3

stout, that it could scarcely have belonged to a leaf less than several feet long. What is shown upon the accompanying plate is a pinna only, of a bipinnate leaf.

53

SPHENOPTERIS? BIFIDA.

Communicated by Mr. Witham, from the mountain lime-stone of the lime quarries of Birdy House, near Edinburgh.

So little has this the appearance of a Fern, that you would say it had surely been a root of some aquatic plant, or at least its submersed stem, with such dissected leaves, as we now find floating in ditches, or pools, and belonging to Myriophyllum, Utricularia, Ranunculus, and the like.

In fact, if we compare it with Utricularia minor, we shall see that in both plants the leaves have the same dichotomous divisions, terminating equally in fine subulate points; nor do we know how fragments of such Utricularias as U. *minor*, *intermedia*, and many others, are with any certainty to be distinguished from many species of Trichomanes and Hymenophyllum. Compare, for instance, U. *intermedia* with Brongniart's figures of several species of Sphenopteris, (*Hist. des veg. foss. t.* 48. *f.* 3.—*t.* 49. *f.* 2, &c.) and you will at once remark the almost perfect identity of outline, division, and venation. Nevertheless, as this is a bipinnated plant, it probably was not an Utricularia, all the known species of which are simple; and it is also not likely to have been a part of any species of the other genera above alluded to. On the contrary, it must, with the imperfect knowledge we possess about it, be arranged in the genus Sphenopteris, in the vicinity of S. myriophylla, from which it is known by its leaves not having more than three or four primary divisions, and these not radiating from a common centre, and repeatedly dichotomous, but arising from a flexuose axis, and simply bifid.

54

SIGILLARIA PACHYDERMA.

Euphorbites vulgaris. *Artis Antediluv. phyt. t.* 15.
Rhytidolepis ocellata. *Sternb. Flor. der Vorwelt, fasc.* 2.
p. 36. *t.* 15.
Sigillaria pachyderma. *Ad. Brongn. Prodr. p.* 65.

A class of Fossils, the larger stems of which occur in great abundance, not only in the various members of the Coal formation proper, but also in many of the beds of the subjacent mountain limestone series.

These stems have often escaped compression, and stand perpendicularly across the strata, sometimes having roots proceeding from them on all sides ; they are generally, if not always, surrounded by an envelope of fine crystalline bituminous Coal, as much as an inch in thickness. The longitudinal flutings which are the characteristic marks of this Fossil, generally are indistinct on the lower part of the larger stems, but this is not always the case.

That these plants have been hollow, and of little substance, is proved by their extreme thinness when horizontal, and by their being frequently composed, when upright, entirely of sand-stone, within the outer coating of Coal. This is often of a nature different from the rock in which they are embedded, and also frequently contains impressions of Ferns or other plants; and the internal layers of sand-stone when separated, instead of being horizontal, present a dished appearance.

Plate 54, represents one of these Fossils found immediately above the Coal in Killingworth Colliery, near Newcastle. It is figured and described in the Transactions of the Natural History Society of Newcastle, vol. 1, page 206, by Mr. Nicholas Wood, who also presented the specimen to the Museum of that Society. The lower part was 2 feet in diameter, coated with coal, and indistinctly fluted; the roots were embedded in shale, and could be traced 4 feet or more from the stem, branching and gradually growing less (one of the larger of these is shewn at fig. 2); these roots, as well as the whole of the stem, were composed of fine grained white sand-stone, totally different from the rock in which the lower portion of the Fossil was enveloped, but agreeing perfectly with a bed surrounding the higher part.

For the purpose of examining this Fossil, Mr. Wood had the stone surrounding it removed, during which operation we had an opportunity of visiting it twice in the mine, and of taking drawings and measurements. At the height of about 10 feet the stem was partially broken and bent over, so as to become horizontal; and here it was considerably distended laterally, and not more than an inch thick, having the flutings comparatively distinct.

This stem formed one of a considerable group, not less than 30 being visible within an area of 50 yards square, some of them larger than this individual, all presenting the same general characters, and appearing to have grown where they now stand. The specimen under review, conveyed the idea of having been able, by the aid of its strong spreading roots, to withstand the force of the current which had prostrated and scattered its weaker congeners. Above the height of 10 feet, however, it had been partially broken, and overthrown, the stem having a south west direction.

The perpendicular trunks of this Fossil are often the cause of serious accidents to the colliers, as the coaly envelope, weakening the cohesion of the strata, causes them to detach themselves, and suddenly slip out of the roof, after the seam of coal has been removed from below, when they leave circular holes, 1 to 3 feet in height, sometimes 4, or even 5 feet in diameter.

Such are the Geological facts connected with Sigillariæ. The next question is, what analogy did they bear to existing plants? According to

Mr. Artis, they were related to Euphorbiaceæ ; in the opinion of Schlotheim, Palms are their kindred. Von Martius refers them to Cacteæ ; Brongniart formerly considered them completely different from any thing at present known ; but now, with Count Sternberg, places them among Filices.

Of these opinions, the only ones that require examination are those of Artis, Von Martius, Von Sternberg, and Brongniart.

The arguments that have been adduced in support of the analogy of Sigillaria with the trunks of Tree Ferns, are not very clear ; they seem to depend more upon a supposed resemblance between the scars left upon the surface of Sigillariæ, and the thick cortical integument that enveloped their trunks, than upon any thing else. The resemblance between the scars of Tree Fern stems and those of Sigillaria, appears to us to be altogether imaginary, for in all the stems of modern Tree Ferns there is uniformly a ragged margin, a spiral arrangement, a denseness of situation, and a size which are wholly at variance with what occurs in Sigillaria ; and as to the presence of a distinct cortical integument, there are two difficulties in the way of admitting that as proof of an analogy between Tree Ferns and Sigillaria, either of which seems to us to be fatal.

Firstly, In Sigillaria this cortical integument overlaid the whole surface of the stem, and the leaves were evidently articulated upon .it (as appears by the cleanness of the scars they have left behind), being connected with the woody, or, at least, central axis, by one or two bundles of vessels that passed through a thick cortical mass. Now, in Tree Ferns, the leaves are not articulated with the stem, leave no clean scar behind when they fall away, and have, for the most part, no woody axis, with which they may be connected by bundles of vessels; but, on the contrary, are mere prolongations of certain sinuous woody plates, which form the hollow cylindrical stem. Secondly, it is plain that the cortical integument of Sigillaria was of the nature of true bark, that is, separable freely, without tearing, from the woody axis, as is evinced by the remains of the decorticated specimens that are so common; while, in Tree Fern stems, the cortical integument is of the nature of that spurious bark in Palms, and other Monocotyledonous trees, which is not more separablefrom the axis than strips of the wood itself.

Another argument against the identity of Sigillariæ and Tree Fern stems, is furnished by M. Brongniart himself, although he does not admit its value. That excellent observer remarked in the Coal mines of Kunzwerk, near Essen, a stem of Sigillaria, the position of which enabled him to follow it nearly its whole extent. The stem was laid parallel to the floor of the gallery, at about the height of the eye of the observer; near the base it was about a foot in diameter, and appeared broken, not terminated naturally. It was, like all stems deposited in the direction of strata, compressed so as to be almost flat. Following this stem in the gallery, he was astonished to find that it extended uninterruptedly to more than 40 feet, its diameter gradually diminishing, so that it was not more than 6 inches across at its upper end. That end, instead of terminating suddenly, was divided into two branches, each about 4 inches in diameter, which diverged for a few inches, and was then interrupted by a partition in the rock. Now this bifurcation, which M. Brongniart considers strongly corroborative of the affinity of Sigillaria and Ferns, is of no value whatever, as an evidence of dichotomous ramification ; because in all cases when a lateral bud developes from a pre-existing axis, or when two terminal buds develope together, bifurcation must obviously be the consequence ; but it will not be dichotomy, unless the developement of terminal buds repeatedly takes place, to the exclusion of all other buds—which was not observed. Besides, we do not know that Tree Ferns would grow in a dichotomous manner, if they were to branch. On the contrary, we know that when they have been seen accidentally to branch, *they are not dichotomous,* as is proved by a plant of Dicksonia arborea now growing in the Garden of the Horticultural Society.

Had the hypothesis just objected to been supported by a Botanist of less knowledge than M. Brongniart, we should have been satisfied to refute it by referring to the figure of a true Tree Fern, Caulopteris primæva, at t. 42 of this work, and

to those of Sigillaria, which immediately follow this article ; but we have thought it due to
M. Brongniart's high reputation to go at length
into the question.

With regard to the opinion of Mr. Artis, that
Sigillaria was related to succulent Euphorbiæ, and
of the learned von Martius, that they may be compared to Cacteæ, there can be no doubt, that, as
far as external characters go, it approached these
plants more nearly than any others now known,
particularly in its soft texture, in its deeply channelled stems, and, what is of more consequence,
in its scars, placed in perpendicular rows between
the furrows. It is also well known that both these
modern tribes, particularly the latter, arrive even
now at a great stature; further, it is extremely
probable, indeed almost certain, that Sigillaria was
a Dicotyledonous plant, for no others at the present day have a *true separable* bark. Nevertheless,
in the total absence of all knowledge of the leaves
and flowers of these ancient trees, we think it better to place the genus among other species, the
affinity of which is at present altogether doubtful.

SIGILLARIA PACHYDERMA.

(*Corticated.*)

———

See t. 54.

———

A portion of the same species as the last, from the shale of Killingworth Colliery.

The surface of this specimen is deprived of its bark, so that the scars which remain are merely the places through which the vessels of the stem passed into the leaves.

It has long since been shewn by Brongniart, that all Sigillarias are to be found in two states; firstly, that in which the bark remains uninjured (*corticated*); in these the scars are clean, broad, and well defined; and, secondly, that in which, the bark being destroyed, nothing remains but the passage through which the vascular system of the leaf communicated with the stem

o

(*decorticated*); in these the scars are narrow, small, sometimes indistinct, and often double, or reniform.

According to Mr. Artis, a further difference in character is to be expected in the same species. He states, that in Euphorbites vulgaris, which seems to be a very perfect state of this Sigillaria pachyderma, the surface of the younger part has the scars in single rows, rhomboidal with a reniform impression near the upper end, with a decurrent line on either side; but that in the same stem when old, every other furrow widens, and becomes concave, while the alternate ones contract, so that between each fluting there is two rows of scars instead of one. As we have not verified this observation, we think it better for the present to consider the double and single rowed specimens as belonging to different species.

56

SIGILLARIA ALTERNANS.

(*Corticated.*)

Syringodendron alternans. *Sternb. flor. der vorw.* 4. *p.* 50.
t. 58. *f.* 2. (*Corticated.*)

From Cramlington Colliery, in Northumberland.
This specimen consisted of the coal itself, and when sliced and examined microscopically, shewed traces of tissue, the nature of which we hope some day to have an opportunity of explaining. Count Sternberg procured it from the Coal mines of Eschweiler.

The double row of approximated oval scars, each with a smaller scar in the centre, is the diagnostic sign of this Fossil.

57

SIGILLARIA RENIFORMIS.

(*Decorticated.*)

Palmacites sulcatus. *Schloth. Petrefaktenkunde. p.* 396. *t.* 16. *f.* 1. (*decorticated.*)

Palmacites canaliculatus. *Ib. t.* 16. *f.* 2. (*decorticated.*)

Sigillaria reniformis. *Ad. Brong. in Ann. des Sc.* 4. *p.* 32. *t.* 2. *f.* 2. (*corticated.*)—*Prodr. p.* 64.

Rhytidolepis cordata. *Sternb. Tent. flor. prim. p.* 23.

? Syringodendron pulchellum. *Sternb. Fl. der vorw. fasc.* 4. *p.* 48. *t.* 52. *f.* 2. (*decorticated.*)

The Coal mines of Eschweiler, Essen, and Waldenburg, in Germany, and of Newcastle, have all produced this Fossil, which also appears from Von Schlotheim to occur in the Greywacké of the Harz, and in the Quadersandstein, near Gotha.

In its corticated state it presents impressions of roundish kidney shaped scars, in the centre of which is a point, and at a little distance on each side a curved mark : when stripped of its bark, as in the accompanying specimen, it has oval contiguous scars, arranged in pairs.

Sternberg's Syringodendron pulchellum, from the Argillaceous Schist of Waldenburg, with the scars of the decorticated specimen a short distance apart, is apparently this same species at a more advanced age.

58

SIGILLARIA CATENULATA.

(*Corticated.*)

? Lepidolepis syringioides. *Sternb. flor. der vorw. fasc.* 3. *p.* 40. *t.* 31. *f.* 2. (*corticated.*)

In Coal from Jarrow Colliery.

Unless this is the Lepidolepis syringioides received by Count Sternberg from the mines of St. Ingbert, it is an undescribed species, remarkable for the appearance of its oval scars, which, touching one another at the ends, form a kind of chain ; the spaces between the furrows are about two inches across.

59

SIGILLARIA OCULATA.

(*Corticated.*)

Palmacites oculatus. *Schloth. Petrefaktenk.* 394. *t.* 17. (*corticated.*)

Syringodendron complanatum. *Sternb. Fl. der vorw. fasc.* 3. *p.* 40. *t.* 31. *f.* 1. (*decorticated.*)

Rhytidolepis. *Cotta Dendrolith.* t. 17.

Sigillaria oculata. *Brongn. Prodr.* 64.

From Killingworth Colliery.

This remarkable species is readily known by its oval scars, which are almost as broad as the spaces between the furrows, and which, having a well defined smaller scar in their midrib, have some faint resemblance to an eye, or to that appearance which naturalists call *ocellated;* it is very nearly the same as Sigillaria notata, which is chiefly distinguished by the form of its scars.

This seems to have been a small species.

60

PECOPTERIS POLYPODIOIDES.

? P. Polypodioides. *Ad. Brongn. Prodr. p.* 57.
? Pecopteris crenifolia. *Phillips' Geol. Yorks. p.* 148. *t.* 8. *f.* 11.

We are indebted to Mr. Bean, of Scarborough, for fine specimens of several very interesting fossil plants from the Yorkshire Coast. They all belong to the Oolitic series, and will be gradually figured by us in this work.

That which is the subject of the present plate is among the most remarkable. In the first place, it is a Fern in which traces of the fructification are extremely well preserved : a case so rare, that Adolphe Brongniart was only acquainted with six instances when he wrote his Prodromus in

P

1828; and, secondly, it has a very striking resemblance to one of the commonest of the recent Ferns found in this island.

It occurs very rarely in the shale of a rich bed of fossil plants at Gristhorpe, near Scarborough, first discovered by Mr. Bean; our specimen lies among fragments of a very narrow Poacites, and of a Pterophyllum. The whole leaf, or frond, seems to have had an oblong outline, and to have been perhaps seven or eight inches in length, of which rather more than three inches now remain. Its rachis is so much destroyed, that nothing can be determined as to its original surface. The leaf was pinnatifid nearly down to the midrib; the segments were nearly linear, about an inch long, with an obtuse termination; each segment is traversed by a strong middle rib, upon which, nearly perpendicularly, are implanted veins, which bifurcate a little beyond the midrib; of this bifurcation one arm is directed obliquely towards the apex of the segment, and is stopped about half way between the midrib and the margin, by a .round spot, which indicates the presence of a *sorus*, or mass of fructification; the other arm again bifurcates, and apparently reaches the margin; at least we thought we distinctly made this out, by holding the specimen in a particular light; we are, however, by no means sure, that its divisions do not stop short of the margin; the fact is, that the veins are so very indistinct, that we have found much difficulty in detecting them at all. The margin of the segments looks, in some places, as if it had been divided into little teeth; but, in others, it is evidently quite entire; and we have no doubt, that the former appearance is only caused by the breaking up of the black carbonaceous matter that has given the impression.

It is evident that our Fossil is referable to Adolphe Brongniart's genus Pecopteris; but as the figures, illustrative of that genus, are not yet published, we have no means of knowing to what species; we conjecture only, that it must be his P. *polypodioides*, from the aptness of the name, and from its having been procured by him from the lower Oolite.

If we compare it with recent Ferns, we cannot fail to be struck with its great resemblance to the very common *Polypodium vulgare;* a plant extremely variable in size, and in the outline of its segments, but in many states scarcely distinguishable from the fossil. In both, the outline of the leaf and of the segments, the arrange ment of the veins, and the situation of the sori, are the same; but in the Fossil the margin of the segments appears to have been entire; while, in the recent species, it is, we believe, always serrated. Beyond this, we really find so little of moment, that we doubt, whether, if a recent Fern were discovered, with so much similarity, and so little discrepancy, it would be considered more than a variety of Polypodium vulgare.

LYCOPODITES FALCATUS.

———

Young and Bird. Geol. Surv. Yorks. t. 2. f. 7.

———

No doubt this is the " Plant, with small round crowded sessile leaves," figured by Messrs. Young and Bird, in their Geological Survey of the Yorkshire Coast. By those Gentlemen it was obtained from the sand-stone of either the Saltwick or Hawkser Cliffs. The beautiful specimen, from which our drawing was made, was sent us by Mr. Bean, from the under shale at Cloughton, where it is of very rare occurrence.

Like the last, this is again an instance of remarkable resemblance between the plants of the Oolitic series, and those of the present day. Without taking any particular species, for that would be difficult, where comparison has to be established between so imperfect a relic as this, and species that, throughout a whole group, are exceedingly similar to each other,—we should say, compare this with such plants as *Lycopodium complanatum,* or any of the same section, and the likeness will be found so extremely strong, as to leave no doubt of their complete analogy.

Such recent plants are, on the one hand, allied to Ferns, with which they agree in the presence of vascular tissue, more or less perfect in their axis, and in their mode of curling up in the nascent state; on the other hand, they resemble mosses, from which they are known by their ramifications, and very different organs of reproduction. Their stems divide by forking repeatedly; and are covered closely with leaves, which are arranged in two rows, having their edges vertical with respect to the axis of growth, not horizontal. These leaves are placed alternately, and are furnished with lateral smaller leaflets, of the nature of stipulæ.

In the Fossil, the figure of the larger leaves is distinctly and strongly falcate, with an obtuse extremity, and with a perfectly entire margin. Of the smaller leaflets but very imperfect traces are to be discovered; they, however, certainly exist, although they are not shewn in the magnified drawing at fig. 2.

Unless this be the Lycopodites Williamsonis,

of Adolphe Brongniart, from the lower Oolite, a species of which neither figure nor description have yet been published, it must be altogether new; for we can meet with no trace of it in any other work than that above referred to. It is one of the prettiest species that have yet been obtained in the rich beds of fossil plants at Scarborough.

TÆNIOPTERIS VITTATA.

Scitaminearum folium. *Sternb. Flor. der Vorw.* 3. *p.* 42. *t.* 37. *f.* 2.

Scolopendrium. *Young and Bird. Geol. Surv. Yorks. t.* 2. *f.* 9.

Tæniopteris vittata. *Ad. Brongn. Prodr. p.* 62. *Hist. des Végét. Foss.* 1. 263. *t.* 82. *f.* 1, 2, 3, 4.

Scolopendrium solitarium. *Phillips' Geol. Yorks. p.* 147. *t.* 8. *f.* 5.

From the shale of the Gristhorpe bed, near Scarborough; communicated by Mr. Bean. It has also been detected at Hoer, in Scania, and at Neuewelt, near Bâle; it is regarded by Ad. Brongniart as one of the most common in the Jurassic formations, and as a species characteristic of his third period of vegetation.

Our figure represents a cast of the upper surface of a leaf, of which the extremity has been
destroyed. It was rather more than five inches
long, of a narrow lanceolate form, terminating
rather abruptly, and unequally at the base. Its
veins are close together, quite perpendicular to
the midrib, and either very simple, or once
forked; its stalk is continuous with the midrib,
and seems to have been smooth.

Any one would naturally be led to consider
this very analogous to the recent *Scolopendrium
officinale*, for its general aspect and mode of
venation are strikingly similar; but Brongniart
has met with a specimen which has traces of
round impressions upon it, which may have
been *sori;* and, if so, this could have been no
Scolopendrium, but must have been more like
some simple-leaved *Aspidium* or *Polypodium*, to
which Brongniart compares it rather than to
Scolopendrium.

63

GLOSSOPTERIS PHILLIPSII.

G. Phillipsii. *Ad. Brongn. Hist. des Végét. foss.* 1. 225. *t.* 61. *bis. f.* 5. *t.* 63. *f.* 2.

Pecopteris longifolia. *Phillips' Geol. Yorks. p.* 189. *t.* 8. *f.* 8.

P. paucifolia. *Id. p.* 148.

Communicated by Mr. Bean, from the shale of the Gristhorpe bed, near Scarborough. We are also indebted to Mr. Dunn, the Vice-President of the Scarborough Philosophical Society, for the accompanying drawings from the pencil of Miss Helen Thornhill.

Neither Brongniart, nor Phillips, appear to have known the real structure of this species, both these accurate observers having seen only leaflets separated from their stalk; they, therefore, took it for a simple leaved Fern and the former compares it with certain species of *Grammitis* and *Acrostichum*.

It was, however, as will now be seen, a plant with four-parted leaves, the stalk of which was continuous with the leaflets; the latter were of a figure, varying in form from linear-lanceolate (fig. 2.) to oval, (fig, 1.), and in length from an inch and a half to four inches. The veins are many times dichotomous, anastomozing into a sort of net-work next the midrib.

What we find very singular in this Fossil is, that the leaflets are four, not five, in number, as is the case with modern Ferns of a similar habit; on this account we are unable to compare it with any recent species. Adolphe Brongniart, indeed, points out simple leaved *Aspidia* as analogous; but, as we have just said, he was unacquainted with the true structure of the plant.

We cannot doubt, that figures 1 and 2, are varieties of the same species; their differences in form are only such as we find in different individuals of the same modern species; and it is not improbable, that the broad leaved form (fig. 1.) may be the barren leaf, while fig. 2 is the fertile leaf.

64

CYCLOPTERIS DIGITATA.

C. digitata. *Ad. Brongn. Hist. des Végét. foss.* 1. 219. *t.* 61. *bis. f.* 2, 3.

Communicated by the Geological Society, from the same locality as Pterophyllum minus, figured in t. 67.

We have no doubt of its being the same as *Cyclopteris digitata*, figured by Brongniart, from Scarborough; but in his specimens the ends of the lobes of the leaf were truncated, and uneven: while in ours, which are, however, very much injured, they appear to be rounded. Brongniart, also, represents his leaf as composed of a single expanded plate; in these specimens it is certainly divided into two or three lobes, as we have represented it.

We confess we have some doubt of this having been a Fern; its texture, and general appearance, together with its irregular lobing, being very much at variance with any modern Ferns that we are acquainted with. It, however, answers well enough to the artificial character of the genus; and it is not worth disturbing its name, unless some better evidence of its nature than we at present possess shall be discovered.

POLYPORITES BOWMANNI.

A single specimen of this very remarkable Fossil was discovered by J. E. Bowman, Esq: of the Court, near Wrexham, among the ejected shale of a Coal-pit, near the entrance of the Vale of Llangollen, in the county of Denbigh.

It was about an inch in diameter, of a deep blue-black colour, with a blue lustre here and there, probably caused by the alumine of the shale having been brought out by exposure to rain. Along with the specimen, we received from its discoverer some extremely useful Notes, of which we have availed ourselves in the following account.

Fig. B. 1. in the annexed plate, represents the Fossil of its natural size, and as it appears to the naked eye, with the exception of its being shown much paler than the original, for the sake of distinctness. Fig. B. 2. is the same, much magnified.

It appears to have been a roundish oval body, flat, and marked externally, near the margin, with numerous zones, which follow the border with tolerable regularity; across these zones run rather close lines, converging towards some common centre. The whole of the middle part is even, and unmarked by lines, except in a few patches, as at *b. c.*, where the cuticle seems to have been removed; in these patches the lines are much more close than those at the margins, and do not converge towards a centre, but have directions that are not in accordance in the different patches, neither do they correspond with the converging lines near the margin. In one or two places, as at *a*, dots, arranged with great regularity, are more or less distinctly indicated. The apparent centre of the specimen, which is a good deal injured, seems, by the direction of the lines, to have scarcely been the organic centre. A portion only of the margin was preserved, and the specimen had no sensible thickness.

In a second specimen, subsequently found by Mr. Bowman, and the only other that has occurred, the principal part of the characters now described were equally found, with the addition, that the margin seemed nearly complete; from which it seems, that the figure of the Fossil must have been what Botanists call roundish-ovate, the base of such a figure being the border where the concentric zones and cross converging lines are situated ; the apex, wanting both these, being smoother, more polished, and irregularly indented with minute depressed dots.

It is a matter of great doubt, whether this really belongs to the vegetable kingdom ; Mr. Bowman remarks, that his second specimen might be taken for the scale of a fish, or of some great Saurian Reptile; and we admit it now, without daring to offer any decided opinion about it, chiefly on account of its resemblance, in some points, to some Cellular* plants of the present æra.

There are certain Fungi belonging to the genera *Boletus, Polyporus, Thelephora, Dædalea,* &c., which attach themselves to their support by one side, projecting forward from it, and increasing by periodical additions to their margin, in consequence of which that part assumes a zoned appearance; when these shrivel, they contract into lines or wrinkles, that form radii, lying across the zones. On their upper surface these Fungi are smooth, or more or less velvety or hairy ; on their under side they are perforated with holes perpendicular to the surface, forming what, in the language of Botanists, is a *hymenium porosum.* It is to these plants that we would compare our Fossil; especially the spots at *a, a,* showing dots arranged methodically, to portions of the *hymenium*

* Cellulares. *Introduction to the Natural System of Botany,* p. 307.

porosum. It may be supposed, that the Fossil shews the upper surface, or the *pileus;* the *hymenium* being prevented by its pores from separating from the shale so as to leave an impression. In that case, *b, c,* will be portions from which the cuticle has been torn, and *a a,* will be still deeper wounds, which, having passed right through the *pileus,* lay bare that portion of the *hymenium porosum,* which was connected with the *pileus.*

For the purpose of ascertaining what the effect would be of compressing a recent Fungus of this description, we took a withered dry specimen of the common *Polyporus versicolor* from off a decayed stump, and having enclosed it in plaster of Paris, we separated the mould so formed, when we obtained such an appearance as is represented at A.; the spaces *a* were, however, only made visible by scraping through the pileus with a sharp penknife.

The principal objection to this Fossil being really a Fungus, analogous to those with which we have compared it, consists in the lines in the spaces *b, c,* fig. B. 2. not being in accordance with the radial lines near the margin. It might, indeed, be supposed, that the former have been caused by the pressure of another Fungus lying in a somewhat different direction; to this, however, several objections will obviously present themselves; or, it may be assumed, that the pileus was composed, internally, of two or three layers,

the organic tissue of which was not in corre-
spondence.

With these very unsatisfactory Notes, we commend our Fossil to the enquiries of our readers;
remarking only, that if it is a Fungus, it is perhaps
the first that has been discovered in the Coal
Flora, and that it may be worth considering
whether the *Carpolithes umbonatus* of Sternberg,
referred with doubt to *Cyclopteris* by Adolphe
Brongniart, may not also be something of a
similar nature.

PTEROPHYLLUM COMPTUM.

Cycadites comptus. *Phillips' Geol. Yorks. p.* 148. *t.* 7. *f.* 20.

Among the rocks of the Oolitic series, appear, for the first time, remains of plants related to a tribe called, by Botanists, Cycadeæ.* In their recent state they are small plants, having a thick, fleshy, roundish, or oblong, or occasionally cylindrical and elongated stem, which is never branched, and which is covered with a hard dry coating of scales, that once were the bases of leaves that have fallen off. Their leaves are of a hard leathery texture, are divided in a pinnated manner, and when young are curled up at the points like those of ferns : their veins are, in all cases, undivided,

* *Introduction to the Natural System of Botany, p.* 245.

and proceed in nearly parallel lines from the base to the apex of the segments. These plants are increased by means of male and female flowers, which are diœcious, and collected in terminal cones, composed of scales, after the manner of a Pine-cone. They inhabit countries having a tropical, or sub-tropical temperature, especially the Cape of Good Hope, the West Indies, and South America, and are capable of enduring the extremity of drought without injury.

Of the Oolitic formation they are the characteristic plants, indicating a climate totally different from that which must have been prevalent when the Coal-measure plants were produced, and, in all probability, by no means unfit for the habitation of man.

The stems of these plants are known, when in a fossil state, by the name of " Birds' Nests ;" their characteristic marks, and the proofs of their analogy with modern species, have been amply explained by Professor Buckland, in the Transactions of the Geological Society.

Their leaves were, when first discovered, mistaken for fern leaves, from which they are known by their pinnated mode of division, *combined with* simple veins, which have the arrangement above described.

No remains of their fructification have hitherto been identified in a satisfactory manner. It is, however, not improbable that the impressions found in the Oolite, having somewhat the appearance of a large flower, one of which is represented by Messrs. Young and Bird, (t. 1. f. 1. and 7.) are fragments of this fossil 'cone, broken transversely. If this be so, the parts called petals, in such fossils, will be the scales of the cone, and the stamens, and pistillum, will be the fractured axis.*

Of the fossil impressions of this tribe, some represent the segments of the leaves, connected with the general midrib only by the middle of their base : these are referred, by Adolphe Brongniart, to his genus Zamia ; all others have the segments connected with the general midrib by the whole of their base; they are not, however, on that account, combined into one other genus ; but as they possess certain well-marked modifications of the veins, they are separated into three genera, distinguished thus :—

1. Vein solitary, forming a thick midrib . . *Cycadites.*
2. Veins numerous, of equal thickness . . *Pterophyllum.*
3. Veins numerous, some thicker than the rest *Nilsonia.*

To the genus Pterophyllum, belongs the subject of the accompanying plate. The specimen is from the shale of the cliff at Gristhorpe, near Scarborough, and was communicated by Mr. Bean.

* We have also been obliged with a cast of one of these Fossils, taken in plaster of Paris by Mr. Williamson, the active and intelligent Curator of the Scarborough Museum.

The leaf appears to have been about 11 inches long, and to have been widest at a short distance below the apex, which is destroyed : at the base it gradually tapers into a stalk. The segments, in the widest part of the leaf, are about an inch long, are rounded at the end, and slightly curved forwards, so as to have a somewhat falcate appearance; they vary in width from three to nearly six lines, and as they approach the base become altogether truncate.

Fɪɢ. 1.

PTEROPHYLLUM MINUS.

———

P. minus. *Ad. Brongn. in Annales des Sciences, vol.* 4. *p.* 219. *t.* 12. *f.* 8. *Prodr. p.* 95.

———

From the Upper Sandstone of the Oolitic rocks at Scarborough. Our specimen is in the collection of the Geological Society.

This species was first detected in a small collection of Fossil Plants found in the Sandstone quarries at Hör, a village to the North of Lund, in Sweden. It is well characterized by its narrow leaves, and short, broad, truncated segments, which are extremely unequal in size. Brongniart represents their margin as absolutely perpendicular to the midrib ; but in this specimen they have an oblique direction towards, what we take to be, the base of the leaf. No trace of veins was left in the grit which received this impression.

Fɪɢ. 2.

PTEROPHYLLUM NILSONI.

Aspleniopteris Nilsoni? *Phillips' Geol. Yorks. p.* 147. *t.* 8. *f.* 4.

Sent by Mr. Bean, from the shale of the Gris-thorpe cliff, near Scarborough.

Except that this is so much smaller, it bears great resemblance to P. *comptum,* from which it may be distinguished by the greatest width of the leaf being near the middle, and by its segments being not only more rounded at the end, but also less falcate.

68

NEUROPTERIS RECENTIOR.

Pecopteris recentior. *Phillips' Geol. Yorks. p.* 148. *t.* 8. *f.* 15.

From the shale at Gristhorpe, near Scarborough; communicated by Mr. Bean.

The genera *Odontopteris* and *Neuropteris* are known from each other by the veins of the former proceeding into the segments directly from their base, without collecting into a distinct midrib, and by the veins of the latter gradually diverging from the midrib as they approach the point of the segments. In general appearance, the two genera are extremely alike, and their species have sometimes a most remarkable resemblance, as the present with *Odontopteris Brardii*, and the next species with *O. crenulata*, both of which are Coal-plants.

This must have been a fern of large size. Its rachis in one part is nearly half an inch in diameter; but it must be observed, that in many fossil ferns that part is much thicker in proportion to the lateral branches from it, and to the size of the whole leaf, than in recent species. It was, probably, tripinnate; the last pinnæ were more than six inches long, and very narrow; the segments attached to the midrib by their whole base, having an oblong falcate figure, seeming to have been blunt, and about half an inch long. The remains of the veins are very indistinct, but seem to have been arranged as represented in our figure.

69

NEUROPTERIS LIGATA.

———

Pecopteris ligata. *Phillips' Geol. Yorks. p.* 148. *t.* 8. *f.* 14.

———

Communicated by Mr. Bean, from the same locality as the last.

The leaf was either bipinnate, or tripinnate; its rachis was slender, and quite in the modern proportion to the lateral branches from it. The pinnæ are so broken, that it is impossible to tell their length, but they seem to have exceeded six inches in the lower part of that portion of the leaf comprehended in our plate. The segments were united to the principal midrib by their whole base, from which they tapered upwards into a falcate lanceolate figure, having its margin distinctly toothed beyond the middle; the length of the segments was rather more than half an inch, and their breadth, at the base, about two and a half, or three lines.

70

SIGILLARIA ORGANUM.

Syringodendron organum. *Sternb. Flora der Vorw. p.* 23. *t.* 13. *f.* 1.

From Jarrow Colliery.

This species, although published by Count Sternberg, is not taken up by Adolphe Brongniart. It differs from S. *pachyderma*, already represented at tab. 54 and 55 of this work, in the scars of the corticated specimen being round, instead of angular, and in those of the decorticated specimens being mere dots, instead of a half circle.

The accompanying figure represents both the surface of the wood, and that of the bark; the latter of which is much thinner than that of S. *pachyderma*.

Very large specimens are, occasionally, met

with; sometimes, as much as 2 or 3 feet in diameter.

Neither in this, nor in any other instance that
we have seen, is there any trace of articulations at regular distances; and, nevertheless, from
the state in which Sigillariæ often occur, one
would be led to expect such a structure; for they
are commonly broken across, as if such were the
case. This was particularly remarked in a groupe
of such stems, which were met with in large
quantities, (as many as 10 in 16 yards,) while
driving a store drift in Jarrow Colliery.

71

SIGILLARIA RENIFORMIS.

(corticated.)

See p. 161.

This plate represents the fossil figured at t. 57, in the state in which it existed before its bark was destroyed.

It will be remarked, that, while the scars upon the decorticated specimen consist of two distinct oval spaces, of a regular figure and size, those upon the outside of the bark had a roundish figure, but slightly indented at the two opposite sides.

Such specimens as this, in which all the sharpness of the angles of the recent plant is completely preserved, shew, in a very satisfactory manner, that they cannot have been long agitated in water before they were deposited; and that,

R 2

if they were originally drifted at all, it can only have been for very inconsiderable distances. In our judgment, they are sufficient alone to destroy the theories of those who fancy that the remains of tropical plants, found embedded in Europe, must have been drifted by currents from equatorial regions.

SIGILLARIA? MONOSTACHYA.

———

Communicated by M. De Cardonnel Lawson,
Esq., from a sand-stone quarry, of the Coal-
formation, at Cramlington, in Northumberland.

This is so like a single rib, or fluting, of a Sigillaria, that it is difficult to believe it can be any
thing else; and yet it is as difficult to understand
how one longitudinal portion of a Sigillaria should
be separated from another, in the way this has
been separated; for not only is there not the
smallest trace of tearing, but the whole specimen stands out in very high relief. The outer
coating is coal; the scars project in pairs, more
than the one-eighth part of an inch from the surface of the fossil.

Along the centre runs a sort of depressed line,
the nature of which is unknown.

73, 74, 75.

FAVULARIA TESSELLATA.

———

Phytolithus tessellatus.   *Steinhauer in Am. Phil. Trans. v. 1.*
    t. 7. f. 2.
? Palmacites variolatus.   *Schloth. petrefakt. t. 15. f. 3. A.*
Sigillaria tessellata.   *Ad. Brongn. Prodr. p. 65.*

Found in the Old Coal-formation.

This curious fossil was first noticed by the
Rev. Henry Steinhauer, in the work above referred
to, where a fragment, in iron-stone, is represented
from Shelf.

What seems to be the same thing, was afterwards figured by Baron von Schlotheim, from the
Coal mines of Essen, in Westphalia, and from
Wettin; but, in both these instances, very indif-
ferent specimens were all that had been met with.

For being able to publish the truly beautiful figures at t. 73 and 74, and for the following description, we are indebted to J. E. Bowman, Esq. of the Court, near Wrexham.

The fossil is of fine-grained Sandstone, and was found in a bed of the same, overlying the Coal strata, at Garthen Colliery, near Ruabon, Denbighshire. The whole was about a yard long, of which this alone was preserved.

It retains, on one side, some of the carbonized vegetable substance, which, also, fills the cavities of many of the scars; it is clearly and beautifully detached from its matrix on three sides, and somewhat flattened, so that a transverse section would be an oval. The rows of scars run longitudinally, or parallel with the axis of the stem, with beautiful regularity, each row being separated by a groove; the rows are narrower, and more strongly marked on the sides, which, from its shape, would appear to have been subjected to the least pressure, or, at the narrow ends of a transverse oval section. The scars in the middle of the areæ, are somewhat club-shaped; the central lobe much elongated, and very various in width, and not so deeply sunk as the shorter lateral ones.

Length of the fossil, nearly 14 inches, slightly tapering upwards.

Widest diameter at the broad end, or base, 5 inches.

Narrowest diameter at the broad end, or base, $3\frac{1}{2}$ inches.

Widest diameter at the narrow end, or apex, $4\frac{1}{2}$ inches.

Narrowest diameter at the narrow end, or apex, $2\frac{1}{4}$ inches; but, here, it has been exposed to some greater additional pressure; and there is an additional irregularity in the surface.

There is no indication of a central woody axis.

It appears to have been the stem of some plant, the leaves of which were placed so close together, that their bases, which were square, were in contact. In the total absence of almost all information beyond that which we have given, it is impossible to offer even a guess as to its probable affinity, further, than that it was Dicotyledonous, with an ultra-tropical constitution.

Possibly, it was allied to Sigillaria, with which Adolphe Brongniart combines it; and this is, in some measure, confirmed by the presence of bark, as is shewn at t. 74. But it does not appear to us advisable to unite it with that genus; on the contrary, we should geologically distinguish this Favularia elegans, and some others, from Sigillaria, by the highly important circumstance of the leaves having been in contact at their base, as is proved by their scars. When growing, the appearance of the two genera must have been very different on that account; for, while Sigillaria had its stem loosely furnished with leaves, after the manner of the common forms of plants of the present day, Favularia must have been a mass of densely imbricated foliage.

This specimen is a further proof, that neither the period which intervened between its removal and final deposit, nor the distance it was drifted, could have been considerable. Its angles are as sharp as if it had been newly gathered.

Tab. 73, is a view of this Fossil, of rather less than half the natural size.

Tab. 74, represents the scars of their natural size. Both these are from the pencil of Mr. Bowman.

Tab. 75, is an old and worn specimen, from the Bensham Coal-seam, in Jarrow Colliery; the principal part of it is decorticated, and has a circular depression in the centre of each scar, instead of the long conical spot, which is found in the same situation on the outside of the perfect bark.

CARDIOCARPON ACUTUM.

Adolphe Brong. Prodr. p. 87.
Sternb. Flora der Vorw. t. 7. *f.* 8 ?

In shale, from the Bensham Coal-seam, in Jarrow Colliery.

Fruits are, as is well known, extremely rare in the old Coal-formation, if we except the fossils called Lepidostrobi; a few specimens, apparently belonging to Monocotyledones, and this genus, Cardiocarpon, being the only others that are mentioned by authors.

The species now represented, occurs, occasionally, in the shale, and always, or, at least, most commonly, in groupes; as is the case in the present instance. This circumstance makes it probable, that they were clustered together, when they were growing on the plant, and that they were either deposited where they grew, or that they had been drifted but a short distance.

Each grain is lenticular, always acute at one end, and sometimes so at the other, but more generally obtuse. The acute end (d) appears to have been the apex, and the obtuse end the base.

The face of the grains exhibits two distinct appearances. In some, there is a slightly elevated line running through the axis, from base to apex, and a little scar placed at the very base, across the elevated line, which is perpendicular to it, and which seems to rise out of it. Others have, distinctly, a circle (c) within the margin, the axis of which is traversed by a line (b), which, at its upper end, has the distinct remains of a small double scar ($a.$) The former appear to be grains seen from the outside; the latter from the inside.

Such being the structure of these grains, as far as they retain any decided characters, we are justified in coming to the following conclusions about them.

They, probably, grew in heads, or dense clusters of some kind.

They were didymous; that is to say, they grew in pairs, applied by their faces, c being the line of their commissure, b the impression of their woody axis, and a the scars caused by the passage of the vessels of the axis into each grain.

They were not adherent to the calyx; for it

is to be presumed, that the little scar, described as existing upon the outside, at the base of the grains, indicates the former presence of a calyx at that place.

Little positive, unfortunately, can be concluded from these data, either as to the analogy of Cardiocarpon with recent genera, or as to the fossil genus to which it must belong.

It was, probably, Dicotyledonous: for, if it had been Monocotyledonous, the grains would have been more likely to adhere by threes, than by pairs. The most striking analogy that occurs to us, is with *Umbelliferæ*; to which, however, it cannot have belonged, if we are right in considering the calyx inferior. Had we not ascertained the character of the inner face of the grains, we might have been induced to suspect some affinity with *Cruciferæ*; but the commissure, and other characters of the inside face, render this impossible. *Stellatæ* might, also, be thought to resemble it, if it were not for the inferior calyx; but, upon the whole, we incline to the belief, that, like many other genera of the Coal æra, it has no very positive modern analogy.

As to the fossil genera to which it may be supposed to belong, we would, in the first place, remark, that it is impossible Cardiocarpon should be the fruit of Lepidodendron, or any other Lycopodiaceous genus, as Adolphe Brongniart has conjectured; this is sufficiently proved, by the didymous structure of the fruit, independently of many other considerations. To what other genera it may belong, we do not feel capable of offering any decided opinion. Supposing it to have fallen from the stem of some species of Asterophyllites; then, one might indulge in the suspicion of that genus having been related to *Callitriche*.

77

CALAMITES APPROXIMATUS.

***

Calamites approximatus. *Sternb. Flora der Vorw. fasc.* 4. *p.* 26.
  Schloth. Petrefakt. p. 399. *Artis Antediluv. Phyt. t.* 4.
  Ad. Brongn. Hist. des Vég. Foss. 1. 133. *t.* 24. *and t.* 15.
  f. 7. 8.
C. interruptus. *Schloth. l. c. p.* 400. *t.* 20. *f.* 2.

***

From the shale of Jarrow Colliery. Mr. Artis
had it from the soft sandstone, in Hober Quarry,
near Wentworth; Von Schlotheim from the Coal
Mines of Manebach, Essen, Saarbruck, and
Wettin; and Adolphe Brongniart, from those
of Alais in the Department of the Gard, of Liége,
of Kilkenny, and of Saint Etienne in the Department of the Loire; and, finally, from the Copper
Mines of Ekaterinebourg, in Russia.

It is thought to be readily known from most
others, by the very close joints of the stem; but

it appears, from the specimen now represented, that this approximation of joints is not universal; on the contrary, those towards the upper end are as distant as in other species.

The bark is rather thick, and very much obscures the furrows of the wood.

The three following forms are recorded by Adolphe Brongniart.

Var. 1. Joints very close, deeply impressed and contracted.

Var. 2. Joints more remote, and less deeply impressed.

Var. 3. A smaller kind, with close joints, and very narrow ribs.

78

CALAMITES

(*With Roots.*)

———

From the Newcastle Coal-field.

Up to this time, we believe that no one has seen what can be certainly considered the roots of a Calamite. As every thing which tends to the elucidation of the nature of this singular genus is highly interesting, we have peculiar satisfaction in, at length, being able to state what they are.

Of the three specimens represented in the accompanying plate, A has the joints of its extremity but little contracted; and, from the base of the lowest articulation but one, there springs an arm, with a descending direction, which is irregularly branched; from the articulation above this, springs another arm of the same nature. These are, most undoubtedly, roots, as is proved by the absence of all trace of symmetry in their mode of ramifying. It, therefore, would seem from this, that the lower end of a Calamite has no contraction of the joints, but, on the contrary, has them pretty regularly elongated. C we take to be another instance of roots; in this, however, the specimen is much less distinctly preserved; but, at B, where the articulations become gradually shorter as we approach the end, we have appearances so similar to the supposed roots of C, that it is difficult not to believe them, also, to be of that nature; and if this be the fact, then it would appear, that the test of the root end of a Calamite has still to be sought for, and that neither the lengthened nor shortened joints are characteristic. But to this we have to recur in the next subject.

We are uncertain to what species to refer these fragments; possibly, they are small specimens of C. *arenaceus.*

79

CALAMITES CANNÆFORMIS.

Calamites cannæformis. *Schloth. Petrefaktenk.* 398. *t.* 20. *f.* 1.
Sternb. Flora der Vorw. fasc. 4. *p.* 26. *Ad. Brongn. Hist.*
des Végét. Foss. 131. *t.* 21.

C. Pseudo-bambusia. *Sternb. Flora der Vorw. t.* 13. *f.* 3. *Artis*
Antediluv. Phytol. t. 6.

───────

This is one of the commonest species, being
found in almost every Coal-field in Europe.

It is readily known by its smooth surface, its
distant furrows, which usually terminate acutely,
and by its usually curved tapering figure.

We have placed the drawing now given of it in
the position which it should have, if the long
cylindrical bodies proceeding from it were leaves,
and the specimen itself the apex of a branch. But

s 2

we are rather inclined to believe it to be the base of a stem, and the cylindrical bodies to be roots; for if we compare it with fig. B., in tab. 78, the resemblance is so great, that we can scarcely fail to recognize it; and it is next to certain, that that fossil is a root end. Besides, it will be remarked, that the tubercles which terminate the ribs of the stem, originate near the points *most remote* from the apparent apex; but it is a constant law in vegetation, that leaves originate from that end of a joint which is *next* to the real apex; and there can be little doubt, that these tubercles, because of their regular arrangement, indicate the seat of rudimentary leaves. If this reasoning be correct, then the accompanying figure is reversed, and it is to be considered the base of the stem of Calamites cannæformis.

From these remarks, one useful conclusion may be drawn; namely, that the position of the tubercles upon the stem of a Calamites, affords the only certain evidence of base and apex; the end at which they are seated, will always be the upper end. This is confirmatory of Adolphe Brongniart's opinion, that those curious rounded ends of Calamites, with contracted joints, and short wide ribs, which are frequently met with in collections, are the bases of stems, and not their upper ends, as Artis, and others, have imagined.

INDEX TO VOL. I.

The Synonymes are printed in *Italics*.

———

INDEX.

INDEX.

INDEX.

INDEX.

Tilling, Printer, Chelsea.

THE

FOSSIL FLORA

OF

GREAT BRITAIN;

OR,

FIGURES AND DESCRIPTIONS

OF THE

VEGETABLE REMAINS FOUND IN A FOSSIL STATE

IN THIS COUNTRY.

BY

JOHN LINDLEY, Ph. D. &c. &c.

PROFESSOR OF BOTANY IN THT UNIVERSITY OF LONDON;

AND

WILLIAM HUTTON, F.G.S. &c.

" Avant de donner un libre cours à notre imagination, il est essentiel de rassembler un plus grand nombre de faits incontestables, dont les consé-quences puissent se déduire d'elles-mêmes."—*Sternberg.*

PART II. OF VOLUME I.

LONDON:

JAMES RIDGWAY, PICCADILLY.

1831-3.

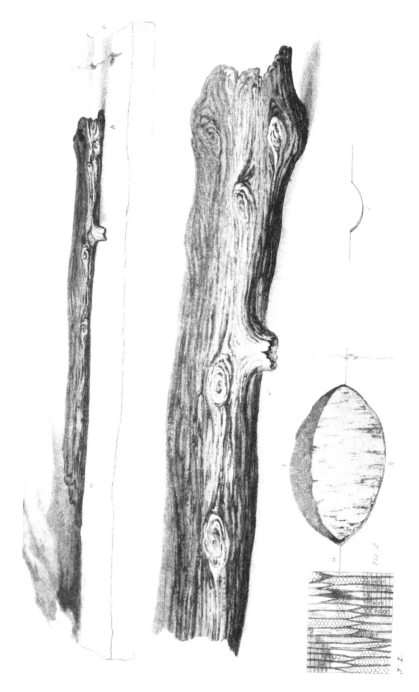

Parker.

Published by J. Ridgway and Sons London July 1. 1832.

Plate 2

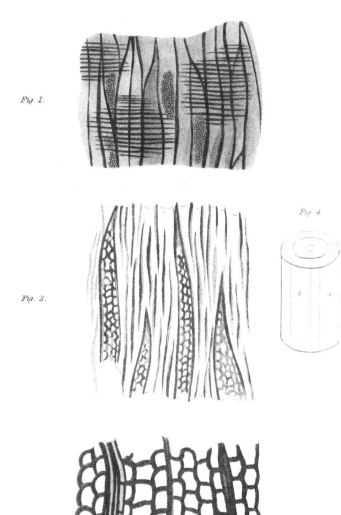

Fig. 1.

Fig. 2.

Fig. 4.

Fig. 3.

Published by J. Ridgway and Sons. London. July 1. 1822.

Plate 3

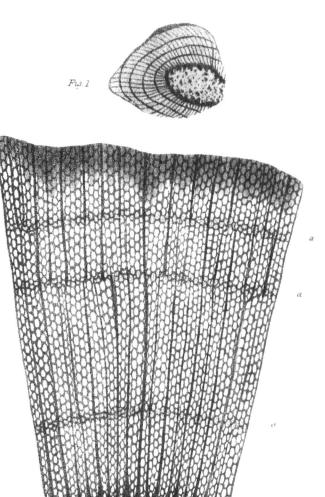

Fig. 1

Fig 2.

Published by J. Ridgway and Sons, London, July 1, 1832.

Plate 4

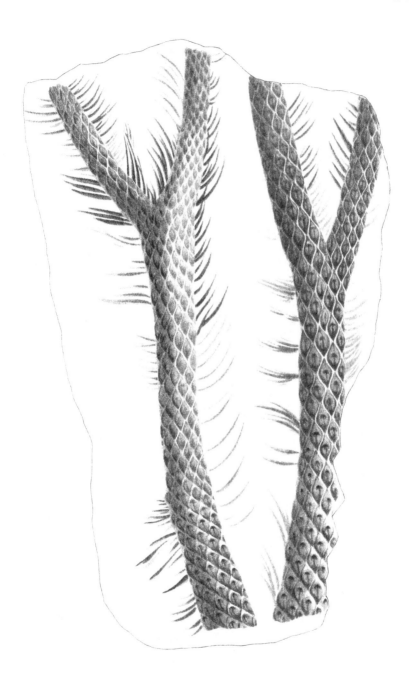

Published by J.Ridgway and Sons London July 1.1831.

Plate 5.

Published by J. Ridgway and Sons, London, July 1, 1831.

Plate 6.

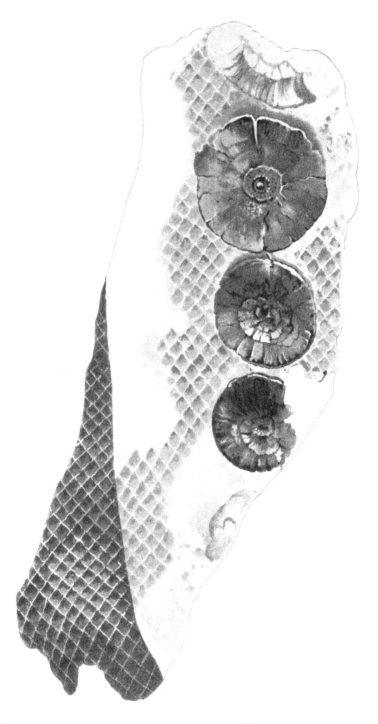

Published by J.Ridgway and Sons, London. July 1.18.

Plate 7

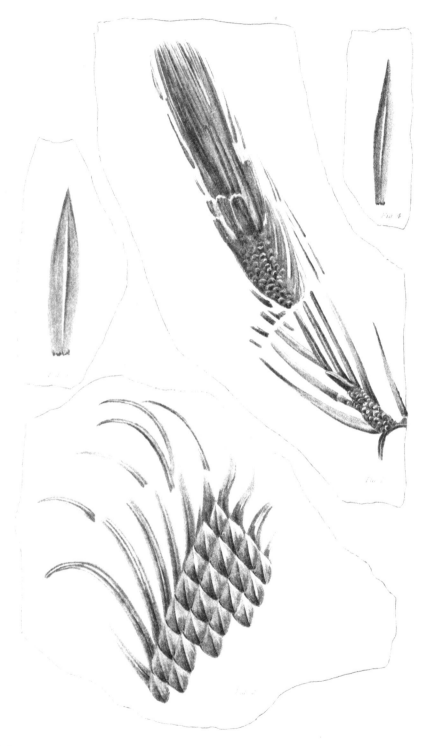

W H

Published by J. Ridgway and Sons London July 1. 1831.

Plate 8.

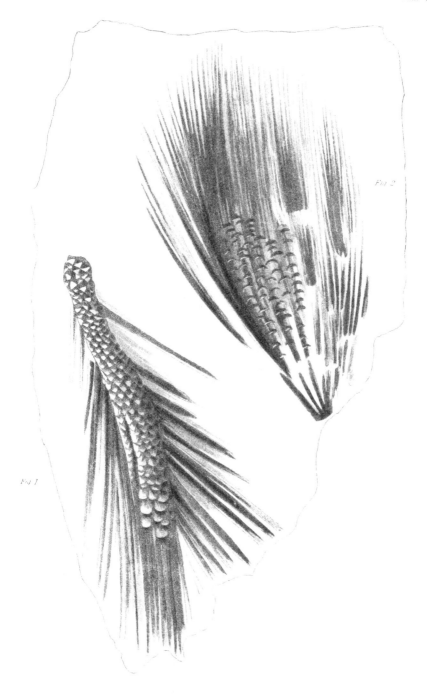

Fig. 2

Fig. 1

W. H

Published by J. Ridgway and Sons London July 1. 1831.

Plate 9

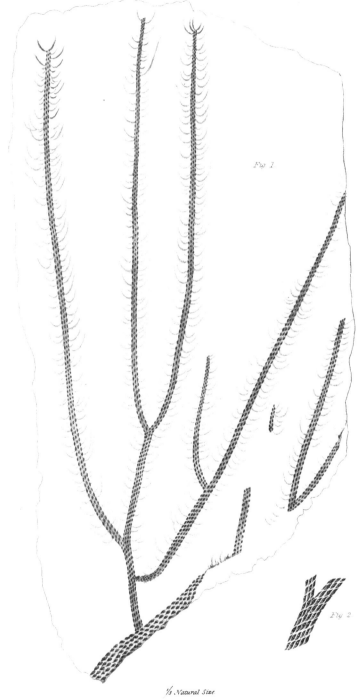

Fig 1

Fig 2

⅛ Natural Size.

W H   Published by J. Ridgway and Sons London July 1. 1831.

Plate 10.

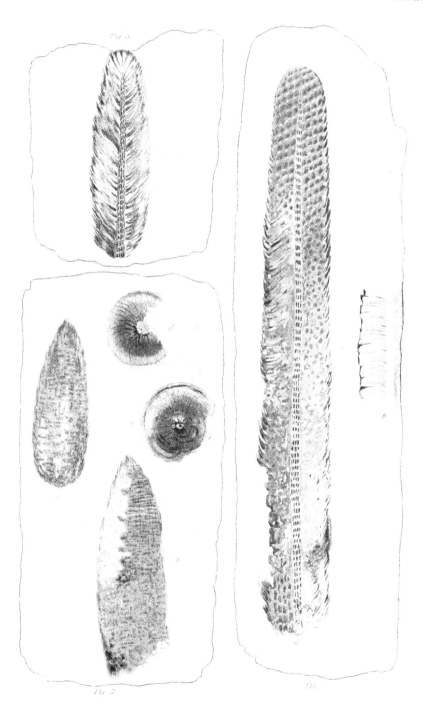

Published by J. Ridgway and Sons, London, July 1, 1831.

Plate 11.

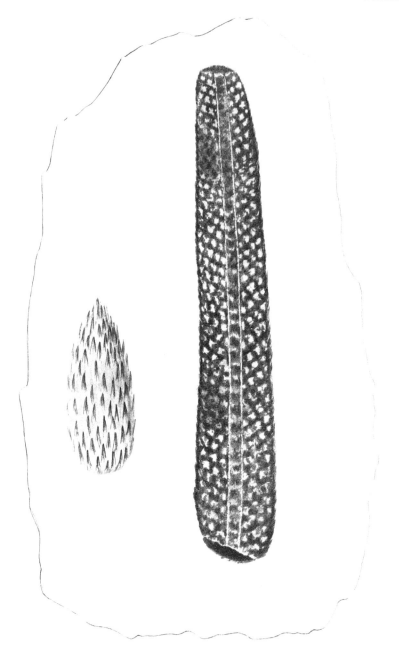

Published by J.Ridgway and Sons London Oct.ʳ 1. 1831.

Plate 12.

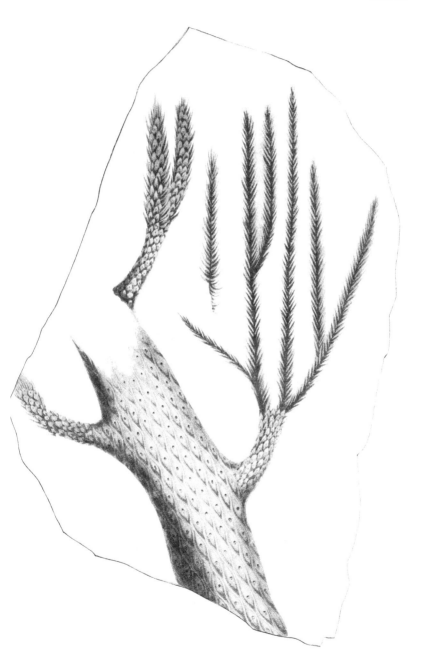

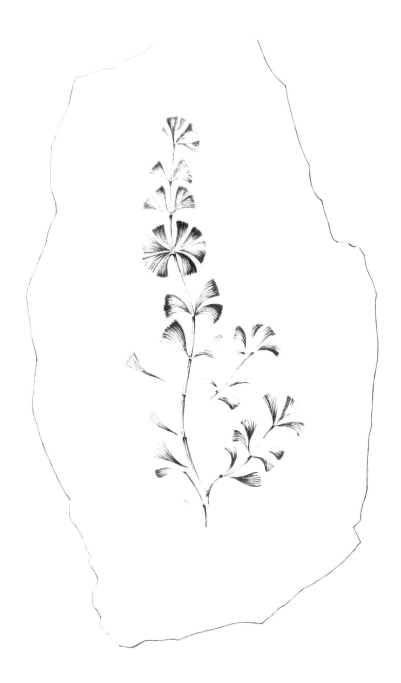

Plate 13

Plate 14.

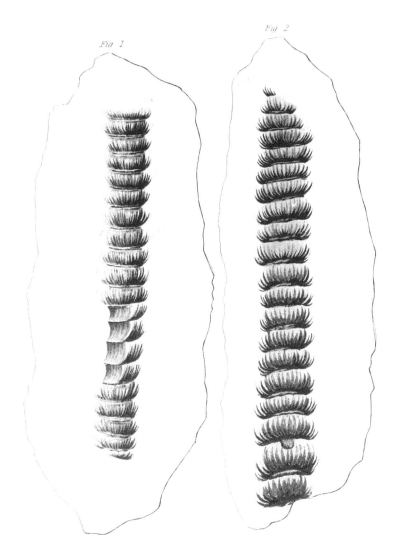

Fig. 1

Fig. 2

Plate 15

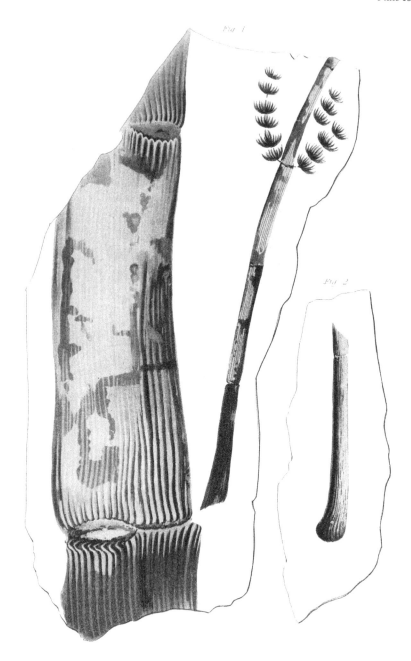

Published by J.Ridgway and Sons London Oct' 1 1831.

Plate 16.

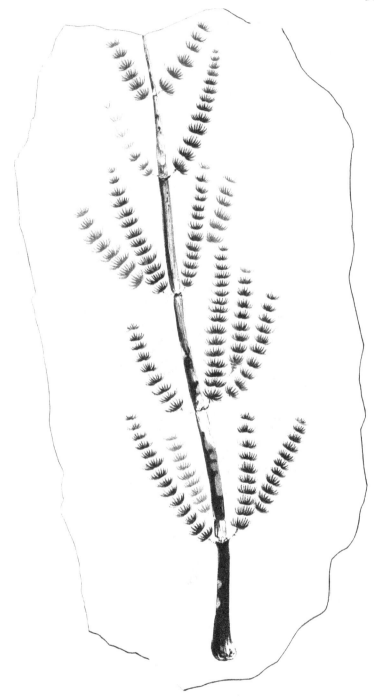

Published by J.Ridgway and Sons London Oct.r 1 1831.

Plate 17

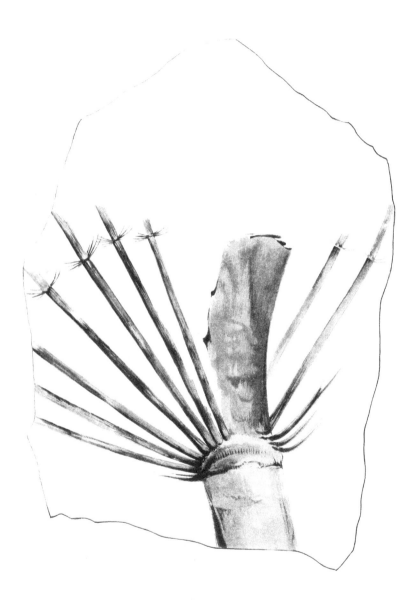

Published by J. Ridgway & Sons. London. 1851.

Plate 18

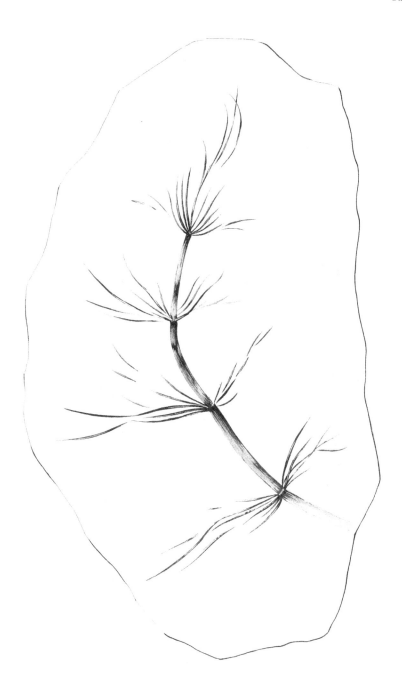

Published by J.Ridgway & Sons, London, 1841.

Plate 19

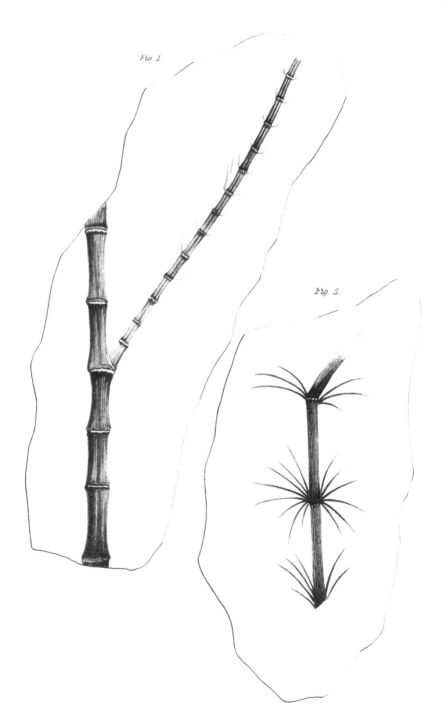

Fig. 1

Fig. 3

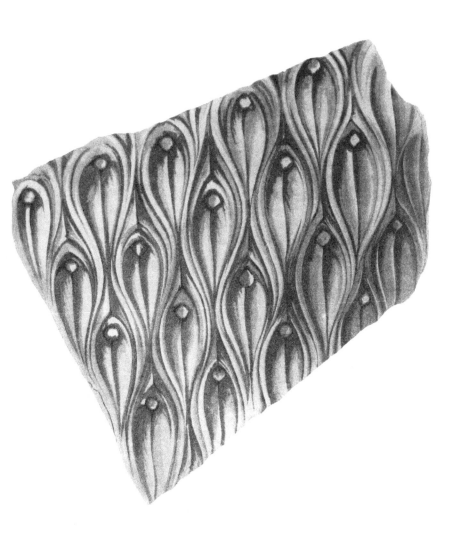

Published by Ridgway & Sons, London. April. 1832.

Plate 20

Fig. 1

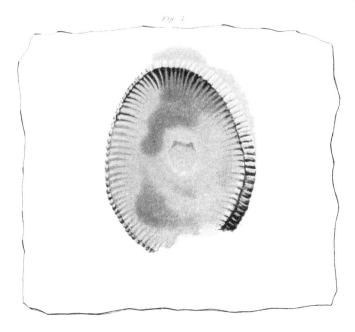

Fig. 2

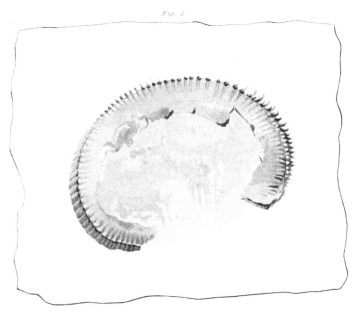

Published by J.Ridgway & Sons, London, 1841.

Plate 21

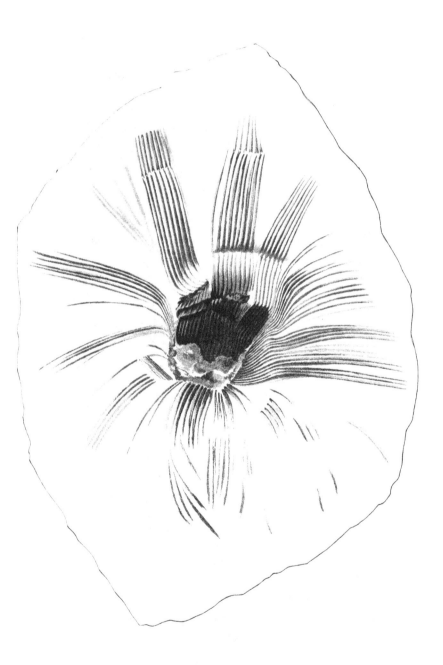

Published by Ridgway & Sons.London. Jan' 1832.

Plate 22.

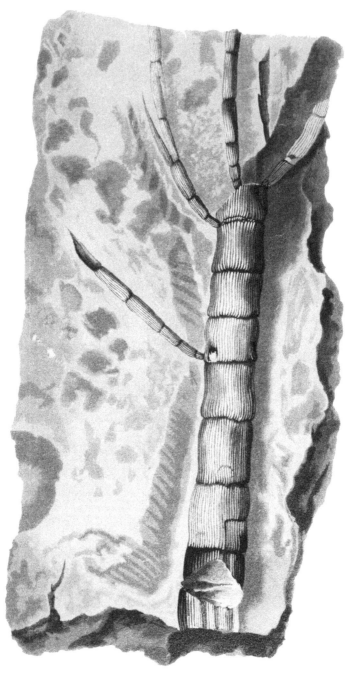

Published by Ridgway & Sons, London, Jan.ʸ 1832.

Plate 23

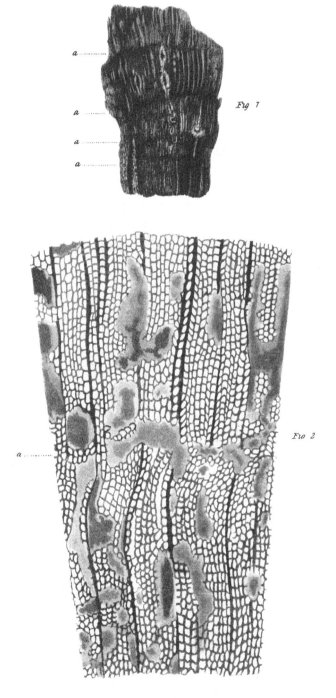

a

a

Fig 1

a

a

a

Fig 2

Pub. by Ridgway & Sons, London. Jan.ʸ 1832

Plate 24

Fig 1

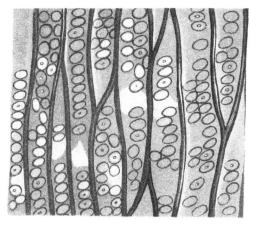

Fig 2

Pub. by Ridgway & Sons. London. Jan.ʳ 1832.

Plate 25

Fig. 1.

Fig. 2.

Pub. by Ridgway & Sons. London. Jan.y 1832.

Fig. 1

Plate 26

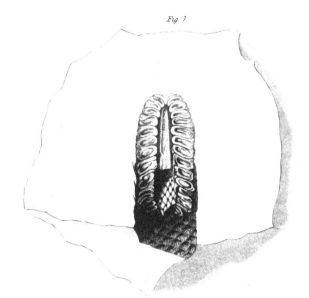

Fig. 2

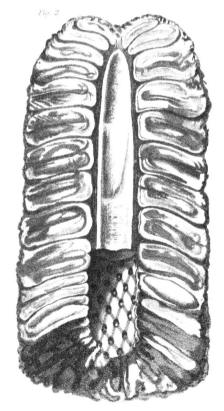

Pub: by Ridgway & Sons, London Jan.ʸ 1832.

Plate 27.

Fig. 7

Fig. 2

Fig. 3.

Pub: by Ridgway & Sons, London, June 1832.

Plate 2

Pub: by Ridgway & Sons London. Jan.' 10.32.

Plate 29.

Pub. by Ridgway & Sons, London. Jan.ᵧ 1832.

Plate 30.

Fig. 1

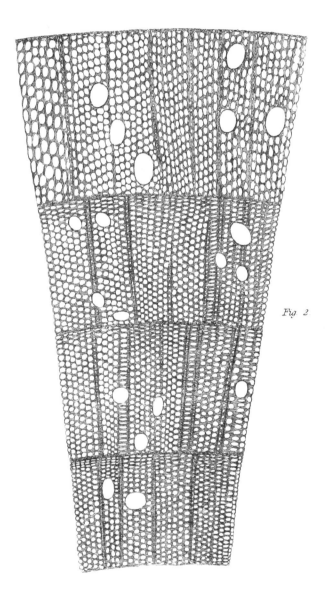

Fig. 2

Pub. by Ridgway & Sons, London, Jan.y 1832.

Plate 31

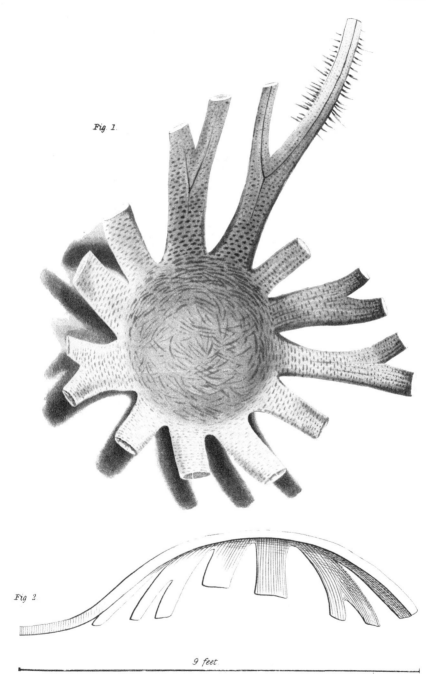

Fig 1.

Fig 2

9 feet

Pub. by Ridgway & Sons, London, Jan.ʸ 1832.

Plate 32

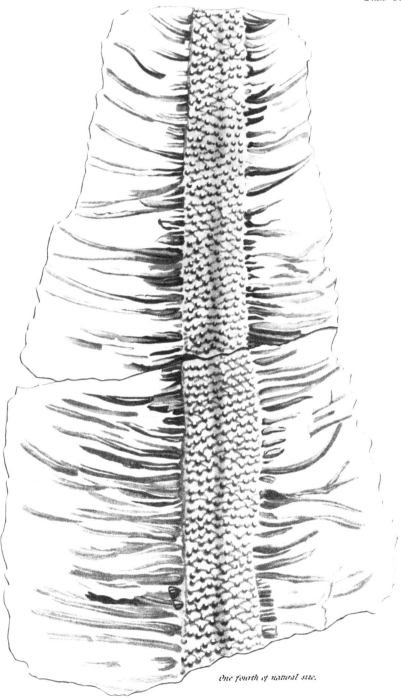

One fourth of natural size.

Pub. by Ridgway & Sons, London. April. 1832.

Plate 33

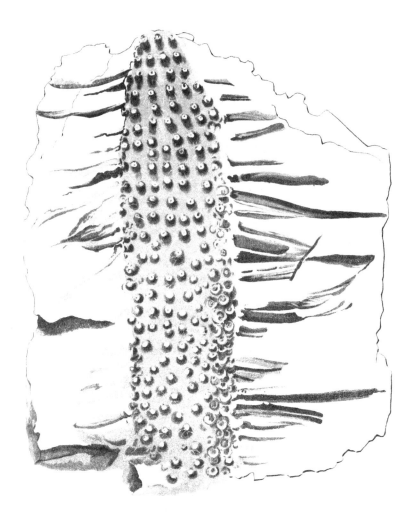

Pub. by Ridgway & Sons, London. April. 1832.

Plate 34.

Pub: by Ridgway & Sons, London, April 1832.

Plate 35

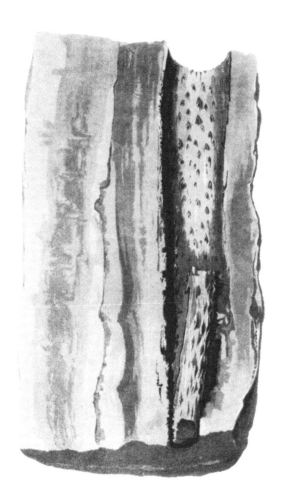

Pub. by Ridgway & Sons. London. April, 1832.

Plate 36.

Pub: by Ridgway & Sons, London, April. 1833.

Plate 37.

Half natural size.

Published by Ridgway & Sons, London, April, 1842.

Plate 38.

Fig. 1

Fig. 2

Pub: by Ridgway & Sons, London. April, 1833.

Plate 39

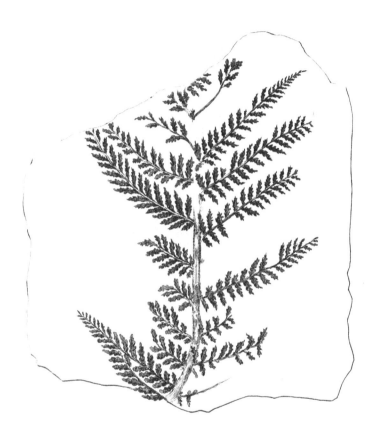

Pub. by Ridgway & Sons, London, April, 1833.

Plate 40.

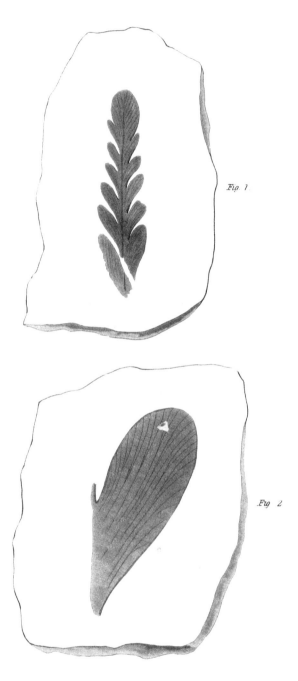

Fig. 1

Fig. 2

Pub. by Ridgway & Sons, London, July, 1832.

Plate 41.

Pub: by Ridgway & Sons, London. July, 1832.

Plate 42

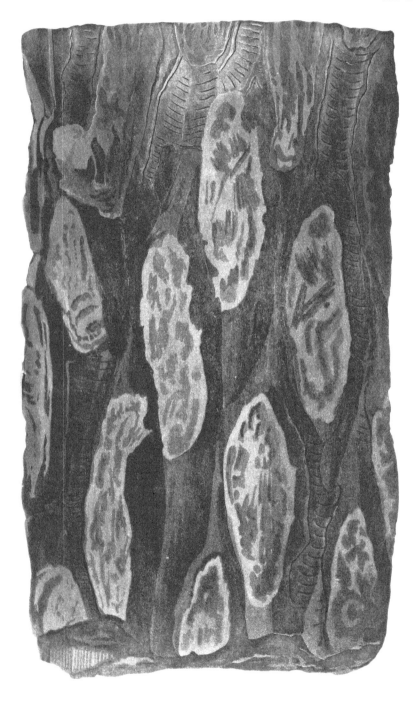

Pub: by Ridgway & Sons, London. July, 1832.

Plate 43

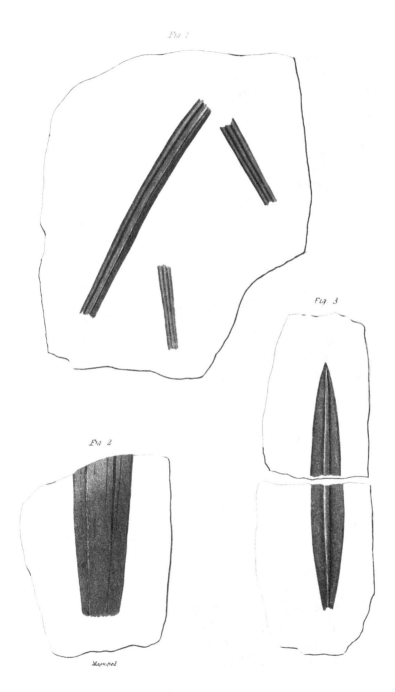

Fig 1

Fig 3

Fig 2

Magnified.

Pub: by Ridgway & Sons, London, July. 1832.

Plate 44.

Pub. by Ridgway & Sons London July 1832.

Plate 45.

Magnified.

Pub. by Ridgway & Sons. London. July. 1832.

Plate 46.

Magnified.

Pub by Ridgway & Sons, London. July, 1832

Plate 47

Magnified.

Pub. by Ridgway & Sons London July 1832.

Plate 48.

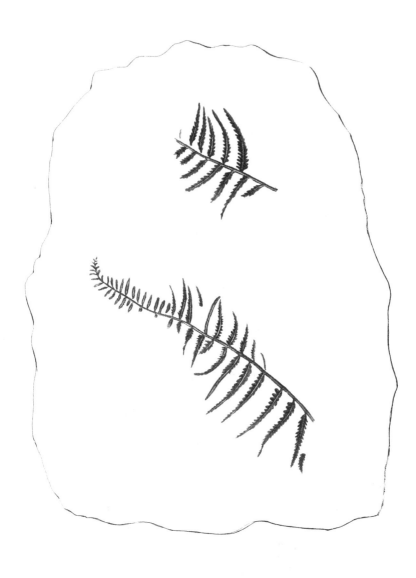

Magnified.

Pub. by Ridgway & Sons, London. July, 1832.

Plate 49.

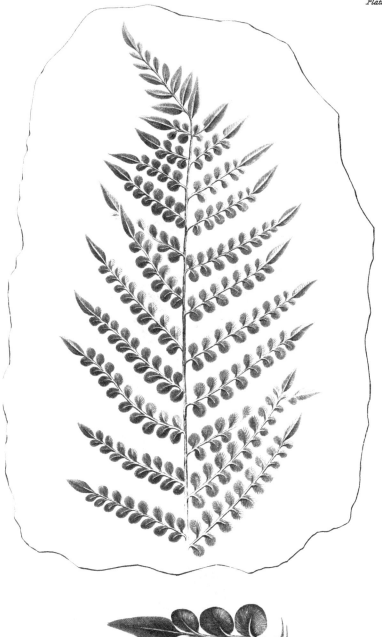

Magnified.

Pub: by Ridgway & Sons. London. July. 1832.

Plate 50

Magnified.

Published by Ridgway & Sons London Oct.ʳ 1832.

Plate 51.

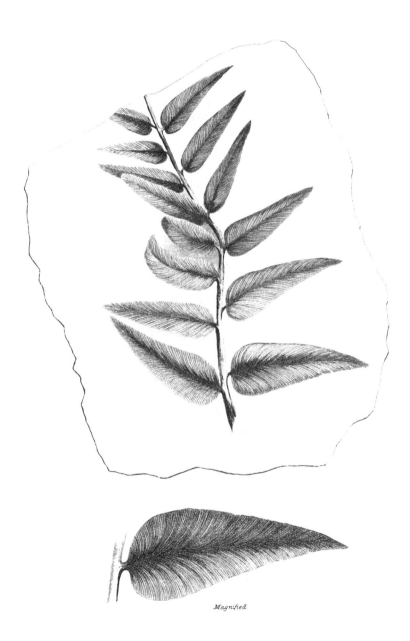

Magnified

Published by Ridgway & Sons London. Oct.ʳ 1832

Plate 52.

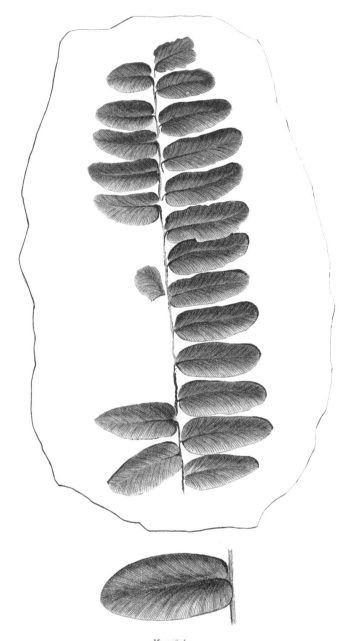

Magnified.

Published by Ridgway & Sons London. Oct.ʳ 1832.

Plate 53

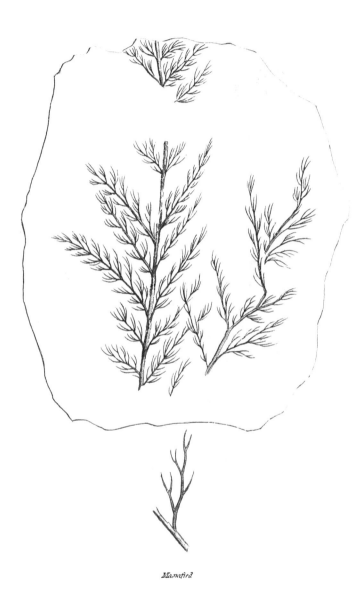

Mamfird

Published by Ridsway & Sons London Oct.ʳ 1832.

Plate 54.

Fig 1

Shale

Fig 2.

Sandstone

12 Feet

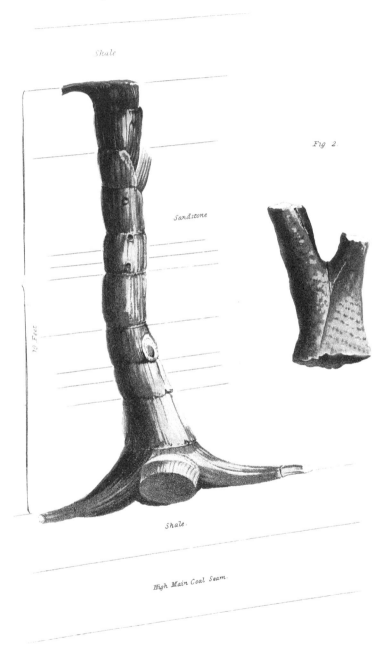

Shale.

High Main Coal Seam.

Published by Ridgway & Sons. London, Oct.ʳ 1832

Plate 55

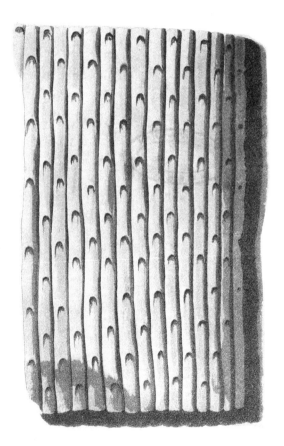

1/4 Natural Size

Publishd by Ridgway & Sons. London. Oct.r 1832.

Plate 56.

½ Natural Size.

Published by Ridgway & Sons, London. Oct.ʳ 1832.

Plate 57.

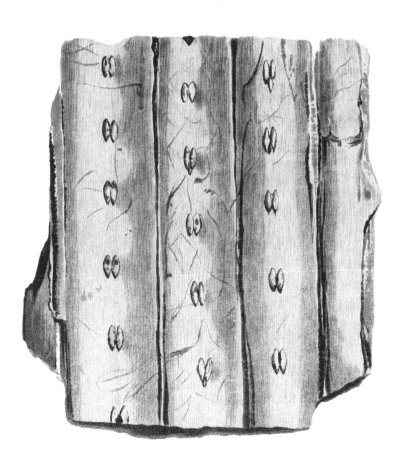

½ Natural Size

Published by Ridgway & Sons, London, Oct.ʳ 1832.

Plate 58.

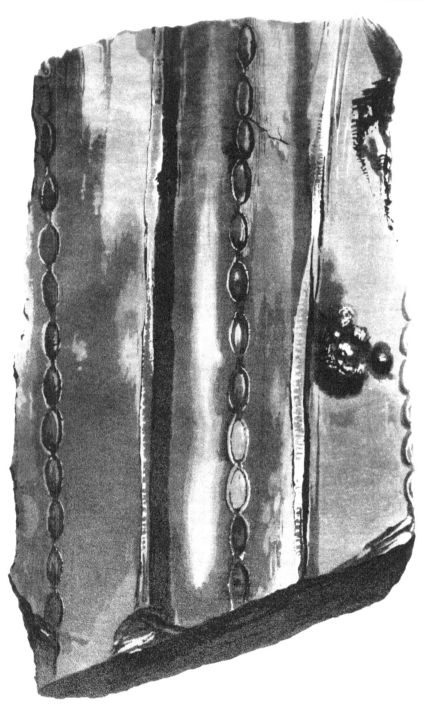

Published by Ridgway & Sons, London. Oct.r 1832.

Plate 59.

¹/₂ _Natural Size._

Published by Ridgway & Sons London. Oct.ʳ 1832.

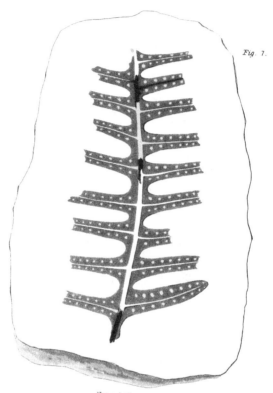

Fig. 7.

Natural Size.

Fig. 2.

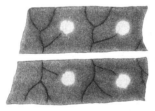

Magnified.

Published by Ridgway & Sons, London, Jan.ʸ 1833.

Plate 61

Fig 1

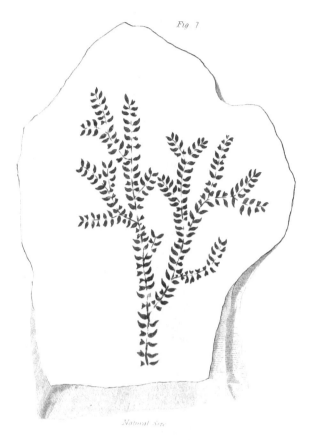

Natural Size

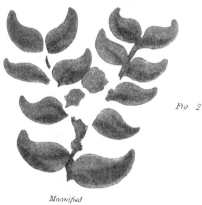

Fig 2

Magnified

Published by Ridgway & Sons, London. Jan.ᵞ 1833.

Fig. 1.

Plate 62.

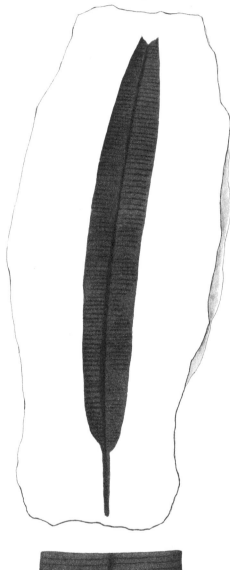

Published by Ridgway & Sons London Jan.y 1833.

Fig. 2.

Magnified.

Plate 63.

Fig. 1.

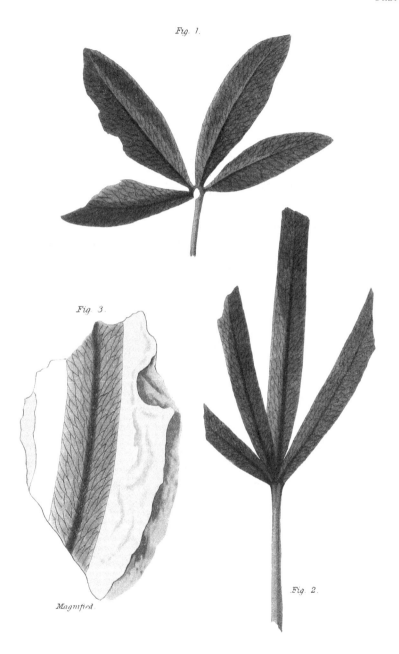

Fig 3.

Fig. 2.

Magnified.

Published by Ridgway & Sons London, Jan.ʸ 1833.

Plate 64

Fig 1

Fig 2.

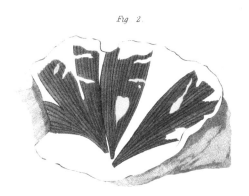

Published by Ridgway & Sons. London. Jan.ʸ 1833.

Plate 65.

B *Fig 2.*

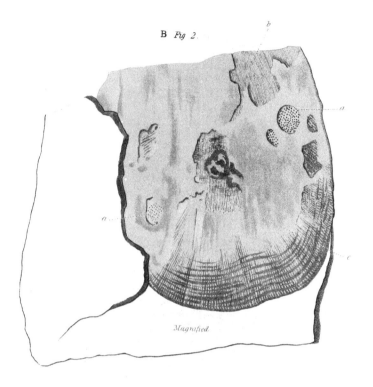

Magnified.

A

B *Fig 1*

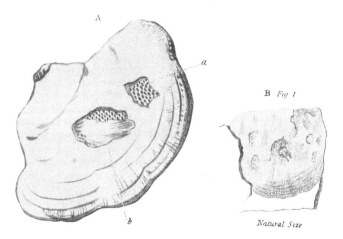

Natural Size

Published by Ridgway & Sons. London. Jan.ʸ 1833.

Plate 66.

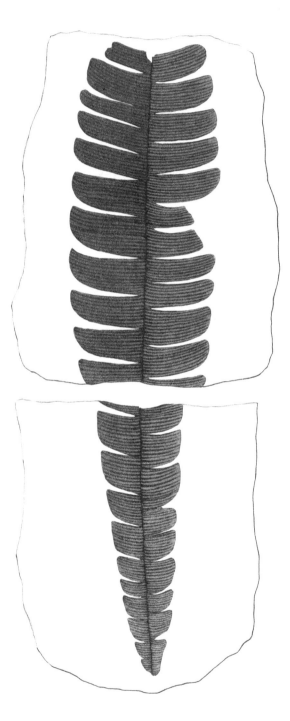

Published by Ridgway & Sons, London, Jan.ʸ 1839.

Plate 67.

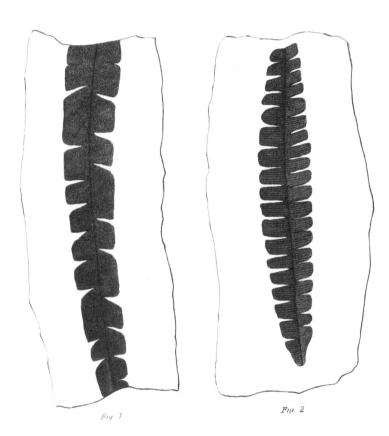

Fig. 1

Fig. 2

Published by Ridgway & Sons, London. Jan.y 1833.

Plate 68

Fig. 1

Fig 2

Mountfied

Published by Ridgway & Sons. London. Jan.ʸ 1833.

Plate 69.

Fig. 1

Fig. 2

Magnified

Published by Ridgway & Son, London, Jan.ʳ 1819

Plate 70.

Published by Ridgway & Sons, London April, 1833.

Plate 71

Published by Ridgway & Sons. London. April. 1833.

Plate 72

Natural Size

Published by Ridgway & Sons, London, April 1833.

Plate 73.

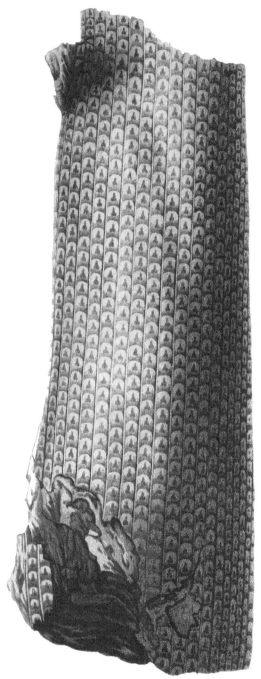

a little exceeding one half the natural size

Published by Ridgway & Sons, London, April, 1833.

Plate 74

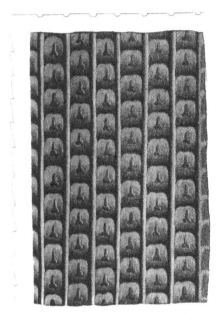

Natural Size of Scales

and intermediate Grooves.

Published by Ridgway & Sons London, April, 1833

Plate 75.

Natural Size

Published by Ridgway & Sons. London. April. 1833.

Plate 76

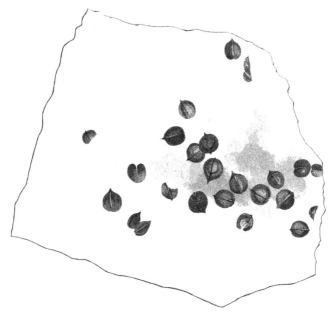

Natural Size.

Magnified.

Published by Ridgway & Sons, London. April 1833.

Plate 77.

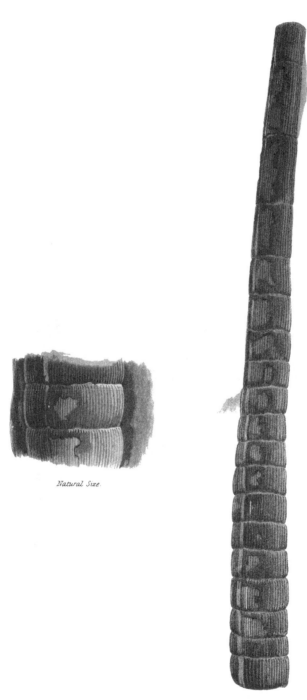

Natural Size.

Published by Ridgway & Sons. London. April, 1833.

Plate 78.

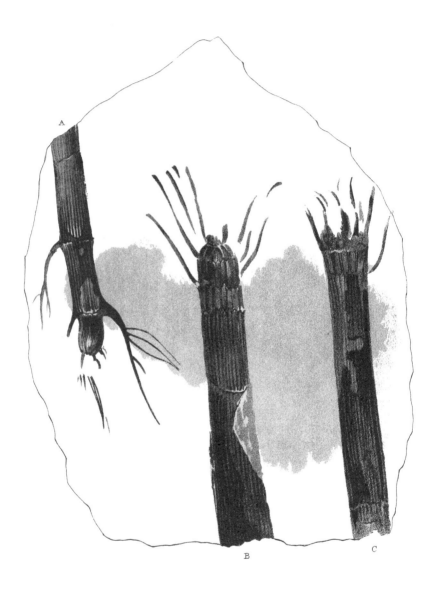

Published by Ridgway & Sons, London. April, 1833.

Plate 79

Publifhed by Ridgway & Sons London. April. 1833

Printed in the United States
By Bookmasters